OBSERVING **EVOLUTION**

OBSERVING **EVOLUTION**

Peppered Moths and the Discovery of Parallel Melanism

Bruce S. Grant

Johns Hopkins University Press
Baltimore

Printed in the United States of America on acid-free paper
9 8 7 6 5 4 3 2 1

Johns Hopkins University Press
2715 North Charles Street
Baltimore, Maryland 21218-4363
www.press.jhu.edu

Library of Congress Cataloging-in-Publication Data

Names: Grant, Bruce S., 1942– author.
Title: Observing evolution : Peppered moths and the discovery of parallel
melanism / Bruce S. Grant
Description: Baltimore, Maryland : Johns Hopkins University Press, [2021] |
Includes bibliographical references and index.
Identifiers: LCCN 2020046103 | ISBN 9781421441658 (hardcover) |
ISBN 9781421441665 (ebook)
Subjects: LCSH: Peppered moth. | Melanism.
Classification: LCC QL561.G6 G73 2021 | DDC 595.78—dc23
LC record available at https://lccn.loc.gov/2020046103

A catalog record for this book is available from the British Library.

Photographs: Unless otherwise credited, all photos are from the author.

To the memory of my mentor,
Professor Donald D. Rabb

Contents

Preface

When I retired in 2001 after teaching in the Biology Department at the College of William & Mary for 33 years, one of my colleagues cheerfully said, "And now you won't have to do any more research," to which I expressed horror. "What? Why on earth would I want to quit doing research?" I should have left it there, but I went on scolding, insisting that (1) the people we hire do research, whether or not they are paid to; (2) they are motivated by curiosity; and (3) they do it because it's fun. It's the best game in town.

Of course, unlike my colleague, I didn't require a cyclotron to carry on. All I needed was a bright light bulb to attract moths.

For the first third of my career, I was a lab-bound geneticist who worked on fruit flies. Then I switched to tiny parasitoid wasps to explore other questions. I didn't fall in love with my research critters until I became a moth trapper in 1983. That conversion didn't happen because I was attracted to moths, in the same way some people find enjoyment in butterflies or birds. I was driven to more deeply explore the phenomenon known as *industrial melanism* (dark coloration evolving in animals living in places affected by industrial pollution), a topic quite familiar to biology students worldwide. Nonetheless, there were still questions to be answered, and new ones to be asked. I became immersed in this research and remain so to this day.

The work took me around the world. In this book I describe my experiences and the remarkable people I met along the way. I tell their stories, including my own daily exploits. But not as a tourist. While travel far from home for extended periods is part of my "Methods & Materials," the book itself is about the scientific discovery of parallel evolution—the rise *and* fall of melanic (black) peppered moths on separate continents. My route to this discovery was a circuitous one, often barking up the wrong tree. Thus this is an adventure story, offering a firsthand account of the history of this research.

While I have intentionally aimed this book at a broad audience, professional evolutionary biologists will, I hope, find it sound and intellectually satisfying. For those especially interested in going beyond my narrative, I have included a substantial bibliography that contains sources for further information. General readers can safely skip the bibliography, in the knowledge that all claims made in the book are supported by extensive, peer-reviewed literature.

Most of all, I hope readers will enjoy learning about how evolution is studied as an observable, ongoing process. It's all around us.

Acknowledgments

Acknowledgments are typically filled with lists of names, along with the various contributions people have made to published projects. My whole book is a means of giving credit to and thanking the individuals who have made the work possible. These are the same people I met along the way while pursuing research with peppered moths. They are the cast of characters—my colleagues, collaborators, and friends. This volume tells their stories. Their names appear in the following pages.

But yes, some people deserve special mention. Jim Murray, in particular, started me on this quest. My wife, Cathy, and our two daughters, Megan and Elspeth, were my field hands, critics, companions, and sources of great joy. They remain in those starring roles.

My long-time colleague and fellow moth trapper, Larry Wiseman, read the complete book manuscript in two days. I took that as a compliment. He has unstintingly provided me with encouragement and advice. Steven Orzack, another dear friend, volunteered to serve as an unofficial literary agent. He recruited fellow evolutionary biologist Richard Gawne as a voluntary reviewer. I was touched by their support and benefited from their constructive criticisms. I also owe thanks for reference assistance to Mary A. B. Sears of the Ernst Mayr Library at Harvard University's Museum of Comparative Zoology.

Most certainly I have failed to mention individuals in this book who have helped me in numerous ways. These include people who permitted me to operate moth traps on their property. Jewel Thomas maintained livestock during my extended absences from Williamsburg and provided photographic services. The esteemed neurologist Lloyd Guth, MD, collaborated on a pilot study assessing altered pH of food plants on larval survival. I apologize to those students who worked in my moth lab but whose names didn't make it into this narrative. They include Leslie Crabtree, Annie Harvilicz,

Kathleen Huffman, Wenda Smith Ribeiro, Heather Scott, and Dan Stimson. Their efforts advanced my understanding of peppered moths.

Anonymous reviewers provided detailed suggestions and spotted my blunders. I am very grateful for the time they have generously invested to help me succeed. I am especially indebted to Tiffany Gasbarrini, senior acquisitions editor at Johns Hopkins University Press, for her welcoming reception of this project and thoughtful guidance throughout. Kathleen Capels taught me what expert copyediting is all about. For her careful scrutiny of the entire production, sincere thanks are owed to production editor Hilary Jacqmin. I am very fortunate to have such talented people in my corner. Whatever errors remain in this book are mine.

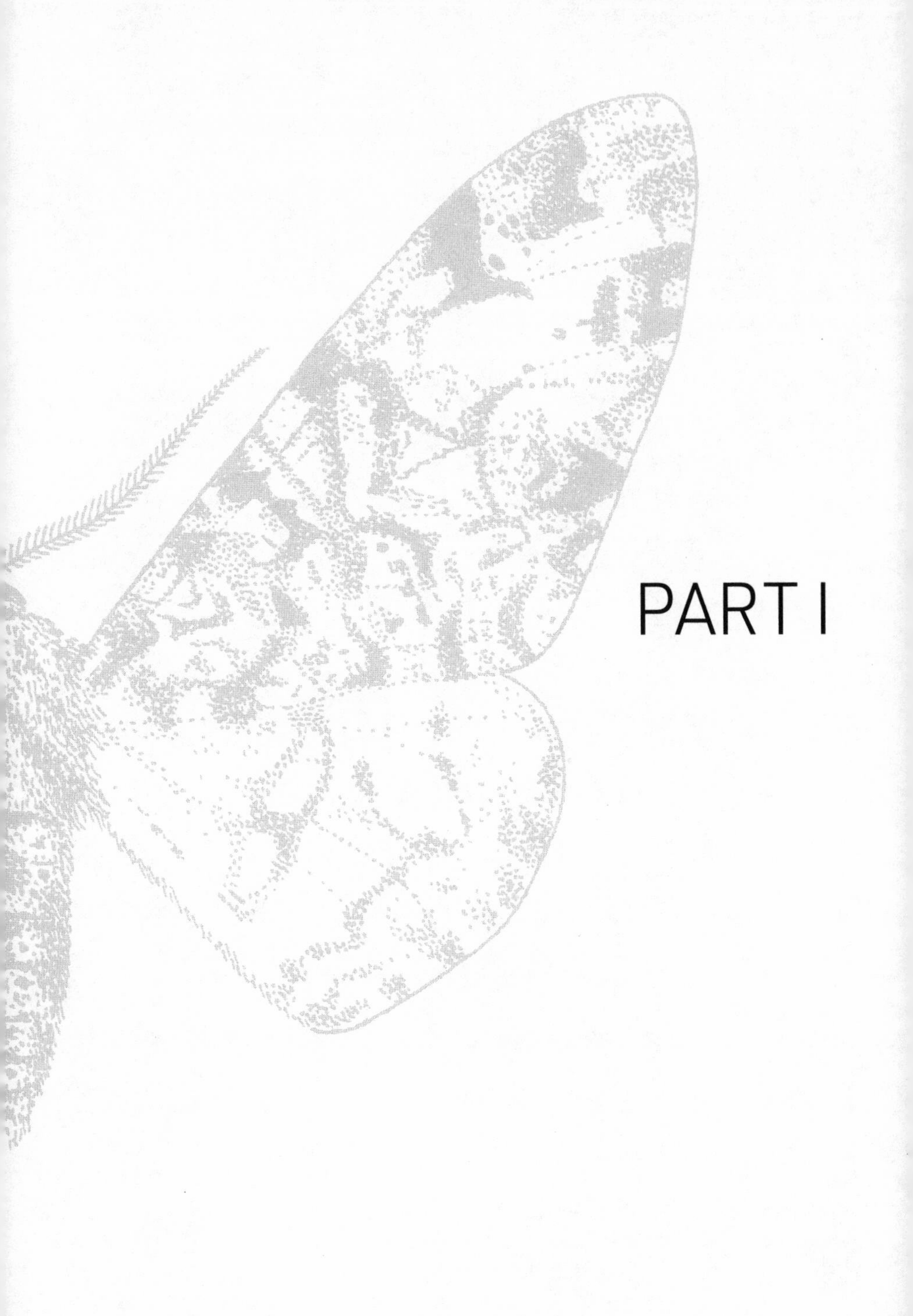

PART I

1 | Passing the Baton

Charles Darwin's *On the Origin of Species* (published in 1859) is, in essence, a book on the origin of adaptations. The process by which adaptations are fashioned—according to Darwin's long argument—is natural selection. During Darwin's own time, natural selection was regarded as a process that was too slow to observe directly, step by step, as it occurred. Evolution was studied as a history, rather than an ongoing process. Times have changed. Natural selection is now witnessed routinely. In some instances, significant changes occur with astonishing speed, such as the evolution of antibiotic resistance in bacteria, or pesticide resistance in insects, or changes in bill dimensions in Galápagos finches. The first and best-known example of observable natural selection traces its beginnings to the decade before Charles Darwin published his most profound book, and this example was explained in full Darwinian terms just 14 years after Darwin's death. The evolution of melanism in *Biston betularia*, commonly called peppered moths, was heralded as "Darwin's missing evidence." I learned about it in school. At the time, I didn't imagine that I'd become personally involved with advancing the field.

My role in the peppered moth chronicles began a century after Darwin's death, when University of Virginia professor Jim Murray traveled from Charlottesville to Williamsburg to talk about his research on land snails. His linked mission was to recruit students to the Mountain Lake Biological Station. Jim was the station's director then, and part of his job was to drum up business for the field courses offered there each summer. Mountain Lake is also a seasonal home to visiting faculty and graduate students who go there to do research in the Appalachians.

Following Jim's lecture, we hosted a reception to chat informally over refreshments. I was eager to bend Jim's ear. Although this was our first meeting, I had learned beforehand that he'd received his PhD from the University of Oxford at the time H. B. D. Kettlewell was in residence. To my

"Did you know him?" Jim replied, "My lab was next door to Kettlewell's. Actually, the two rooms were originally one, but a very thin (and not at all soundproof) partition had divided them. I could hear every word of the bellowing from next door."

Until his death in 1979, Bernard Kettlewell had been the undisputed leading figure in peppered moth research, clearly the Mister *Biston betularia* of his time. But certain features of Kettlewell's work troubled me. I unloaded some of this on Jim as he attempted to nibble on crackers and cheese. I had yet to lay eyes on a living peppered moth, although, in the courses I taught, I routinely included the famous case of industrial melanism to illustrate natural selection. It was, after all, the reigning classic example of evolution in action.

Still, there remained unresolved questions regarding the resting behavior of peppered moths. As with most moth species, the adults are active at night, flying about under cover of darkness, but they spend their daytime hours immobile, apparently hiding from predators. According to Kettlewell, peppered moths, when offered choices of backgrounds on which to rest, nonrandomly select those that most closely match their own appearance, with darker-colored moths predominantly settling on dark backgrounds, and paler moths using lighter-colored surfaces. The mechanism by which these moths accomplish this matching behavior remained untested, at least to the satisfaction of critics. So, as Jim was politely sipping wine, I became the loudest critic in the room and demanded, "Why doesn't someone just do those experiments over again, and this time do them right?"

"Why don't you do them?" Jim said softly, stretching the pronoun "you" for emphasis.

It was more challenge than question, causing my panache to evaporate. In truth, I had toyed with that very idea some years before but saw no easy way to approach the problem. Indeed, I even discussed the possibility with G. R. "Jack" Brooks, one of my early colleagues in the Biology Department at William & Mary, who had taught evolution prior to my arrival there in 1968. Jack wasn't very encouraging, telling me I was just a Podunk guy at a Podunk place. I'd hoped he was only half right. My laboratory experience with insects had been limited to *Drosophila* species (fruit flies) and the parasitoid wasp *Nasonia*, so learning the husbandry of peppered moths would require extramural guidance.

"Easier said than done," I protested to Jim, complaining that I didn't have a ready supply of peppered moths, nor did I know how to breed them. If I wanted to do this work, I'd have to move to England.

"Not necessarily," said Jim. "We have *Biston* right here in Virginia. Dave West catches them every summer at Mountain Lake. If you want to work on them, write up a research proposal and send it to me. Maybe we can even get you a little money."

Jim's recruitment trip was successful.

2 | Peppered Moths 101

In a formal letter dated March 30, 1983, Jim Murray wrote, "I am happy to inform you that your application for a post-doctoral fellowship at Mountain Lake for the summer of 1983 has been approved. The award carries a stipend." The letter went on briefly to explain details about room and board, the waiver of lab fees, and so on, but those initial words were all I needed to read before racing down the hallway to invite my colleagues for beer that afternoon at Paul's Deli, a favorite watering hole for Williamsburg locals. For us, shared celebration was an important component of academic life.

In the months before going to Mountain Lake to begin my work on living peppered moths, I devoted myself to reading about them. Years earlier, I had digested the usual textbook treatments, along with several of the major papers and reviews necessary to teach industrial melanism to college students. But to do research requires becoming intimately familiar with everything already known about a subject, whatever that topic is. Becoming a master of the literature doesn't make one an expert, but it is a necessary first step.

American peppered moths, also known by the common name pepper-and-salt geometers, are a subspecies of British peppered moths. A subspecies is a biological race that has been assigned a formal name by taxonomists. The species, named *Biston betularia*, is divided into geographically separated races, and *Biston betularia cognataria* lives in North America. The species is widely distributed across the Northern Hemisphere and is common in cooler deciduous forests, where its larvae (caterpillars) feed on a wide variety of leaves. Taxonomists once regarded the American and European subspecies as separate biological species, as geographic isolation normally prevents interbreeding between subspecies. They have since changed their minds, because moths from these continents are easily hybridized in captivity, producing fertile offspring of both sexes.

Most of the attention this species has received has come from work on British populations, so much so that some people refer to *Biston betularia* as "British peppered moths." After I'd been working with peppered moths on both sides of the Atlantic for several years, one of my distinguished colleagues asked me how peppered moths got into America. I suggested that perhaps the question should be reversed: "How did peppered moths get into England?" There is speculation on the geographic origin of this species and how it spread elsewhere, but to date there is no hard evidence to support which continent can claim ownership. It's likely that the Bering land bridge played a role between North America and Europe, and perhaps Sakhalin Island and the Kurile Islands were routes to Japan. Finding the mitochondrial Eve (mitochondria are DNA-containing subcellular components that are only inherited from mothers) of peppered moths might help sort out the phylogeography (the historical distribution) of *B. betularia*. Still, most of what we know about peppered moths in connection with the phenomenon called industrial melanism comes to us from England, the birthplace of the Industrial Revolution.

The bare bones of the story are widely known to those who have had a proper biology course within the past several decades. Until about 1848, peppered moths lived up to their common name. What are referred to as "typical" adults are peppered with white and black scales over their wings and body, but on closer examination, distinct patterns of lines mark the fore- and hindwings and are characteristic of the species. Variations are known, and the most conspicuous one is fully melanic, meaning it appears to be solid black, because excess black scales obliterate the normal patterning. Intermediates also occur. These various forms (called phenotypes by geneticists) have been given names by lepidopterists (who study moths and butterflies), as is their custom for polymorphic (having more than one known form) species. The full melanics are called *carbonaria*, the several grades of intermediates are collectively lumped into *insularia*, and the typical peppered form is *typica*, although the word "typical" is usually employed. The abbreviation "f." distinguishes a form name from a subspecies name when both are included. For example, the normal phenotype of American peppered moths is *Biston betularia cognataria* f. *typica*. In practice we toss these names around with much less formality, but it is important to understand that these forms (or phenotypes) occur together within the same populations—indeed, within the same broods of siblings, much like people of different blood types or eye colors occur together within families.

Peppered moth populations across Britain were monomorphic (only one form) before the middle of the 19th century. The peppered form was not only characteristic of this species, it was the only form known at that time. The discovery of a black specimen near Manchester came as a surprise and was worthy of description in a note by R. S. Edleston in 1864, although earlier unpublished records exist, and recent molecular data push the date to circa 1819. By the turn of the 20th century, peppered moth populations in that vicinity were approaching monomorphism for the black form. It was the *typica* form that had become rare (but kept its name as the "type," or first scientifically described specimen). The nearly complete reversal of the percentages of these phenotypes in moth populations was regarded as astonishing, especially as peppered moths complete just one generation per year at that latitude.

Elsewhere in Britain, typicals remained the common form. Indeed, in some regions, black forms were entirely absent, or at least very rare. Closer scrutiny revealed a pattern in the geographic distribution of the frequencies of melanic peppered moths found within populations across the whole of Britain. Around industrial centers, melanic moths were found at higher percentages than typicals within these populations, whereas, well away from these zones, the typicals remained at higher percentages than the melanic forms. The correlation of the incidence of melanism within moth populations to the proximity of those populations to human heavy-manufacturing centers was dubbed "industrial melanism."

The phenomenon is not restricted to peppered moths, as it has been described in numerous moth species, as well as in other kinds of animals, particularly those that apparently hide from predators that locate their prey visually. Peppered moths remain by far the most thoroughly studied case, however, so this account will focus on that work and on the people who have done it. Note that progress in any field is propelled or retarded by human ambition.

Among the initial attempts to explain industrial melanism, the one offered by J. W. Tutt in 1896 is, except for minor details, still taught today. We might have forgotten about Tutt's historical role in framing the problem, as he barely received mention in this regard by his successors. Fortunately, R. J. "Sam" Berry reminded us by publishing (in 1990) an edited excerpt from Tutt's book, *British Moths*. It is worth repeating:

> The speckled *A.* [*Amphidasis* = *Biston*] *betularia*, as it rests on trunks in our woods is not at all conspicuous and looks like a natural splash or scar, or a piece of lichen and this is its usual appearance and manner of protecting itself. But near our large towns where there are factories and where vast quantities of soot are day by day poured out from countless chimneys, falling and polluting the atmosphere with noxious vapours and gases, this Peppered Moth has during the last fifty years undergone a remarkable change. The white has entirely disappeared, and the wings have become totally black, so black that it has obtained the . . . [form name *carbonaria*]. As the manufacturing centres have spread more and more, so the [*carbonaria*] form of the Peppered Moth has spread at the same time and in the same districts. . . .
>
> In our woods in the south the trunks are pale and the moth has a fair chance of escape, but put the Peppered Moth with its white ground colour on a black tree trunk, and what would happen? It would, as you say, be very conspicuous, and would fall a prey to the first bird that spied it out. But some of these Peppered Moths have more black about them than others, and you can easily understand that the blacker they are the nearer they will be to the colour of the tree trunk, and the greater will become the difficulty of detecting them. So it really is; the paler ones the birds eat, the darker ones escape. But then, if the parents are the darkest of their race, the children will tend to be like them, but inasmuch as the search by birds becomes keener, only the very blackest will be likely to escape. Year after year this has gone on, and the selection has been carried to such an extent by Nature that no real black and white Peppered Moths are found in these districts, but only the black kind. This blackening we call "melanism," and the Peppered Moth is by no means the only kind of insect in which this melanic change has been brought about in recent times. . . . But, of course, only those species whose habit it is to hide on fences, trees, stones, etc., in such districts, *i.e.*, on surfaces which are blackened by smoke and damp, are liable to the changes which we have just mentioned. (pp. 305–307)

Unfortunately, Tutt's clear explanation inspired little work on this topic. By the 1920s, the eminent theoretician and a founding father of population genetics, J. B. S. Haldane, calculated the fitness coefficient (a measure comparing the relative contribution of typical and melanic moths to the next generation) necessary to bring about the observed changes in peppered moth populations by natural selection. Others at that time disputed this Darwinian interpretation. Namely, J. W. Heslop Harrison reported that feeding the moth larvae leaves that were contaminated with pollutants induced mutations in adult color development. Harrison's claims did not withstand rigorous inspection, however, as others failed to corroborate his findings in independent attempts to replicate his experiments. Still, marginal contention recurs, as we shall examine later on. More to the point, the phenotypic induction of melanism in adults has never been demonstrated in peppered moths, while controlled genetic hybridization crosses have repeatedly confirmed Mendelian inheritance (traits that are transmitted from parents to their progeny in specific patterns). Note that here we are discussing the phenotypes of *adult* moths. (Body color among caterpillars is altogether different and will be discussed separately.)

Kettlewell appeared on the scene in the 1950s, a half century after Tutt's hypothesis was all but forgotten. In an homage to Tutt published in 1997, the late Denis Owen sharply criticized Kettlewell for not crediting Tutt as the original (or, at least, earlier) author of the explanation for industrial melanism. Owen speculated unsympathetically about the reasons for this egregious error of omission, whether intentional or otherwise. Now that this historical oversight has been disclosed, we may fairly examine Kettlewell's contributions.

To say Kettlewell's part of the story was huge would be an understatement. While he did not establish the concept of industrial melanism, he certainly resurrected it. From the 1920s until Kettlewell began his experiments in the middle 1950s, no substantial research in this field had been done. Tutt himself had conducted no experiments to test his predation hypothesis—that birds ate light and dark phenotypes of peppered moths selectively—so that momentous task fell to Kettlewell. Indeed, even quantitative records about the incidence of melanism in moth populations throughout Britain were haphazard. Thus the organization of massive nationwide surveys also fell to Kettlewell. His classic experiments testing the selective predation hypothesis, and the voluminous field data he compiled from population surveys are presented in a long series of papers (published during the 1950s and 1960s), culminating in his definitive book, released in 1973. Kettlewell's name became synonymous with industrial melanism. His work was featured in virtually every biology textbook worldwide and was standard fare in illustrating natural selection. This is generally still true, but newer examples keep surfacing. Thus competition for space and the fashion for what's "hot" have been squeezing out the peppered moth story in recent years. That's as it should be. Evolutionary biology would be in mighty bad shape if we didn't have other observable examples of natural selection in action. Nevertheless, industrial melanism in peppered moths remains one of the best documented and easiest to understand cases, despite relentless campaigns by creationists to impeach it.

By the time I arrived at the Mountain Lake Biological Station in July 1983, I had read everything Kettlewell had written on the subject, twice—and some more than twice. By that point I had also gone through most of the papers by other workers in the field. Still, I had yet to meet my first peppered moth, face to face. That introduction was made by David A. West. Had it not been for him, I wouldn't have recognized a peppered moth if one landed on my nose.

3 | Catching Moths Using Light Traps

I had been to Mountain Lake previously, but for only a few days to attend a SEEGG (pronounced "siege") meeting. (The acronym for the Southeastern Ecological Genetics Group was later rearranged in some creative way to spell "seepage.") SEEGG meetings were informal, informative, and always enjoyable. The idea of spending a summer session at Mountain Lake was appealing, and it was also embraced by my two daughters (then ages 16 and 8), who eagerly volunteered to serve as my field assistants. My wife found the prospect of having the house empty for five weeks to be a positive, so she could remodel the place without obstructive advice. So, with the family station wagon packed to the roof rack, my daughters and I left the heat and humidity of Williamsburg in search of moths at Mountain Lake.

The biological station itself is two miles up the road from the privately owned Mountain Lake Lodge (where the movie *Dirty Dancing* was filmed). The lodge sat on the shoreline of what was once the highest natural lake east of the Mississippi. The lake mysteriously began to lose water circa 2005, but when we were there, it was nearly 50 acres of clear, cold water, 100 feet deep. The station, owned and operated by the University of Virginia, has no lake, but it does have a large manmade pond, called Riopel Dam, and lots of surrounding woods and streams. It is an isolated and enchanting place. At nearly 4,000 feet in elevation, the weather there is generally gorgeous in the summer, with warm days and cool nights. Except when it rains and the wind blows. The station today is much bigger (in terms of modern buildings) than it was in 1983, but its mission to provide summer field courses and host visiting researchers remains intact. An assortment of cabins, a dining hall, and a large stone laboratory building composed the campus. My daughters and I were assigned the grand Burns cabin, complete with fireplace, as our home for that first summer, and I, much to my satisfaction, was provided with a spacious lab—all to myself—in the main building.

Upon our arrival, I checked in with Jim Murray and thanked him for arranging such wonderful accommodations and making us feel so welcome. He showed me around the grounds and introduced me to a few people along the way. By late afternoon, Dave West arrived from nearby Blacksburg, where he was then on the faculty at Virginia Tech. From the back of his vehicle he unloaded the largest moth trap I had ever seen. I had encountered Dave previously, at meetings, but didn't yet know him well. I had corresponded with him about my coming to Mountain Lake, and I asked him about his experiences with peppered moths. He sent me a reprint of his paper about melanism in *Biston* in the central Appalachians that had been published in the British journal *Heredity* in 1977. He attached a characteristically modest note to the reprint, stating that it contained all he knew about peppered moths. Dave had been a frequent member of the summer faculty at Mountain Lake, and he was there again in 1983 to teach a course entitled Ecological Genetics. He used the moth trap as part of his field lab exercises, and he kept a running record of melanism at the station, dating back to 1968.

Dave explained that his moth trap was fashioned after the type used by Cyril Clarke and Philip Sheppard in England, but somehow it ended up being a bit larger than their model. The basic design is a Robinson trap, named after the inventors (two brothers), but its designation has become a general description, as moth trap designs have been modified to suit the needs of their users. The common feature is a light source at the mouth of a funnel. Moths that pass through the funnel are trapped inside a container until someone lets them out. The light sources for most traps are mercury vapor lamps, used because they are very bright and attract moths from a greater distance than ordinary incandescent bulbs. All bright lamps do not work equally well, as wavelength is a factor. Ultraviolet (or black light) is also effective. There's a bit of trial and error in finding the right bulb for the job, and each moth trapper has his own tales to tell on that score, like fishermen and their favorite lures. There is art to this science.

What attracts moths to lights in the first place? This should seem to be a straightforward question, for which—by now—one might expect a clear answer. Amazingly, the question still sparks debate both inside and outside moth circles. Mobile animals typically respond to stimuli or environmental cues in predictable ways. Some of the simplest behaviors are categorized as *kinesis*, *taxis*, and *transverse orientation*. If a stimulus simply increases or decreases the rate of activity or movement, the behavior is kinetic (undirected movement). Still, such simple behaviors can accomplish great things, and they appear to involve selection or intention. For example, if heat stimulates an animal to move in random directions, it continues to keep moving until it encounters cooler conditions. Such critters are then more likely to be found resting in cool locations, such as in shade or under logs, as if drawn there by a particular attraction.

To qualify as taxis, the movement must be focused either toward or away from the source of stimulation. Animals drawn to a light source exhibit positive phototaxis; if they move away from the light, they display negative phototaxis. In a gradient of light, an animal may move into an optimum level of illumination. It is therefore capable of showing both positive and negative phototaxis, depending on prevailing conditions. The proverbial "moth drawn to a flame" seems by casual inspection to be displaying positive phototaxis, which, in this instance, is clearly a destructive behavior. Moreover, it would seem that whatever is drawing a moth to flame is perhaps the same behavior that would attract a moth to a moth trap: phototaxis. But does phototaxis really explain this behavior? Despite the dubious tale that Kettlewell's sponsor at Oxford, E. B. "Henry" Ford, rented a hot air balloon to look for moths heading upward toward the moon, moths use the moon in other ways.

Dorsal light reaction, a transverse orientation, is a behavior easily confused with phototaxis. It's how some animals know up from down. Certain species use gravity for this and have mechanisms to determine the direction of its pull. Others use light. For example, brine shrimp (*Artemia*) exhibit a response that can easily be demonstrated by shining a light on some swimming shrimp. They always orient their ventral (belly) body surface at a right angle to the light source, so you can get them to swim upside down by moving the light to the bottom of their aquarium. You can see it for yourself by doing this simple experiment at home. Brine shrimp also show phototaxis and will swim into or out of a beam of light, depending on its intensity. Notice the orientation of the body axis to the source of illumination. Phototaxis may draw them to or near the surface of water when the light comes from above, but swimming with the ventral surface up is a transverse orientation.

Are transverse orientations involved in attracting moths to lights? It seems so. Most moth species (but not all) are nocturnal, actively flying at night and resting (or hiding from predators) by day. They are "light inhibited," meaning that they remain motionless once the sun comes up but will readjust their resting positions if in direct sunlight. At night, the only natural lights moths normally see are the stars and the moon, and moths apparently use these cues for transverse orientations when flying. To envision how this works, recall riding as a passenger in a car on a moonlit night. The moon seems to keep up with you, no matter how fast the vehicle is moving. While the angle between the car and the moon is changing as the vehicle is driven along the road, the moon is so far away that the shift is imperceptible. To a moth on a short flight, lights in the night sky stand still, thus serving as reliable cues for transverse orientation.

What happens when a moth encounters an artificial light, such as a street lamp or a moth trap? Unlike flying by moonlight or with illumination from the stars, now the angle between it and the artificial light source changes acutely as the moth flies by. It is behaviorally programmed to hold the angle constant, so it readjusts its flight pattern and body axis and goes into orbit around the light. Orbits are seldom perfect, so if the orbit expands, increasing the distance between the moth and the light source, the moth may escape its attraction and move away. If the orbit shrinks a bit, however, the angle will change even more rapidly, forcing increasingly faster corrections on the part of the moth. Ultimately, a moth very near the lamp is experiencing angular changes so quickly that it goes completely haywire trying to make corrections. It starts flying erratically all around the bulb, unable to escape as it tries to achieve a constant angle. Moth traps employ baffles around the bulbs to interfere with the frenzied flights of moths at the mouth of the funnel. Simply put, the moths crash into the baffles and fall through the funnel. Once inside, still bathed in light coming through the transparent funnel, the moths encounter textured surfaces—strips of cardboard or egg cartons—they can cling to. Once the moths have landed, the angle from the light source stops changing, and they are no longer stimulated to readjust that angle. Now they are light inhibited, and they stay put for the rest of the night. Well, some of them do. Others bat around inside the trap all night, seemingly ignorant about what they are supposed to do.

You don't need a moth trap to observe this behavior. Just switch on a porch light on a warm summer night and watch how the moths gather around, with some circling the bulb in frenzy, while others are sitting quite still on the ceiling or wall near the bulb, having surrendered to "daytime" mode.

By late afternoon we had Dave's giant Robinson trap set up on the bank of Riopel Dam, near the edge of the woods. We strung extension cords through the trees and over the road to reach Burns cabin, providing electrical power to operate the mercury vapor lamp. As darkness fell, we switched on the light bulb and waited for the action to begin. Earlier, Jim Murray came by to cheer us on, and he lamented that there was no practical way to cover or turn off the street light in the parking lot at the other end of the dam, about a hundred meters away. Competing lights near a moth trap cut down the size of the catch, as moths become distracted by other sources of illumination. When a lone moth trap is the sole source of light on a warm night, the catch will be huge.

A Robinson trap should not be opened at night, or the moths inside will escape. This means once the trap is set up and switched on, the worker is free to go away until early the next morning. Taking the rest of the night off, however, is not a requirement. A lot of trappers periodically check on the trap during night to see how it's doing. It's reassuring to watch the cloud of night-flying insects whirling above the device. One can even look through the clear plastic funnel to see what's already inside the trap. With practice, it's easy to identify familiar species at a glance. This acquired habit requires shielding one's eyes from the glare of the mercury vapor lamp, as it is painfully bright, and its emissions in the ultraviolet range can damage the retina. Still, moth workers routinely hold their hands up to block the direct rays and squint into traps, just for the fun of it. Night watching is optional.

By sunrise, most nocturnal moth species will have "clamped down" for the day. Those inside a moth trap accomplish this by clinging to the cardboard placed inside for that purpose, although they will clamp onto just about any surface that is sufficiently textured. At this point it is safe to open the trap without losing the inmates. Actually, there are lots of species that are a bit too jumpy to be caught by this method, so when the big funnel is removed from the top of the trap, they simply fly away. Fortunately, many other species are more cooperative and remain clamped down for all they're worth. Probably one reason peppered moths have been favorite subjects for study is because they fall into the latter group of cooperative moths that are hard to disturb once they have clamped down. Plus, they are robust and easy to identify.

Before actually opening the trap, it is a good idea to look around its exterior and the immediate surroundings for moths that have clamped down outside the trap. This must be done before birds show up and pick them off. Birds quickly learn that a moth trap is a breakfast smorgasbord. The early bird is a baneful bird.

Once the trap is opened and the skittish moths have bolted, those remaining may number in the dozens, the hundreds, or very much higher, depending on the previous night's conditions. If the trap is not overly crowded with moths or overrun with swarms of other species, such as caddisflies, it's fairly easy and fun to sort through the night's catch. Peppered moths may or may not be present, but there are always a lot of other interesting moths to examine, each species with its own evolutionary history that is begging to be explored. The variety represented and the intricate beauty otherwise unavailable for such easy inspection make every day like Christmas during moth season. At the time I began moth trapping, I had little appreciation for any of this, but fortunately I was guided by an expert. David West reached into the moth trap and picked up a piece of cardboard. "Aha!" he announced in his rich basso cantante, "Here is your *Biston*."

Indeed, it was the very first *Biston* moth of the thousands I would collect over the next quarter century of fieldwork. This moth species was so famous, and I had looked forward to this day with such great anticipation, that I felt as though organ music should be playing, or trumpets sounding, or perhaps I should run back to Burns cabin and fetch my bagpipe. Instead, I got out my camera and took a photo of the catch's star moth—and Dave's thumb.

4 | Camouflage

Unlike butterflies, which are easily spotted during the day as they flutter about sporting vivid colors that oftentimes advertise noxiousness to predators, most moth species are drab by comparison—and conspicuously absent. Kettlewell reminded us that Charles Darwin was fooled by the apparent paucity of moths when visiting the tropics at a locale where Kettlewell later caught more moths than at any other place in the world. Kettlewell had benefit of a moth trap. Darwin didn't. Without modern sampling equipment, we'd hardly even know of the existence of many familiar moth species. We see them in swarms by lamps at night. But where are they during the day?

Moth species that are both drab and small are lumped together as "LBTs" (little brown things) and routinely ignored by amateur collectors. The larger, more charismatic macrolepidoptera grab all the attention. Size counts. And the diversity among them is bewildering. With practice, common specimens can be assigned to at least some of the major families, such as the Geometridae, Lymantridae, Noctuidae, Notodontidae, Saturniidae, and Sphingidae. Professional lepidopterists recognize a couple dozen families, into which thousands of species are assigned, based on phylogenetic descent (the evolutionary history and relationships of species or other groups). But even the most casual observer is struck by the resemblance of so many moth species to tree bark, or twigs, or leaves, or lichens, or moss, or stones, or flower petals, or fungus, or even bird droppings. We can easily spot such moths in a moth trap, or when they remain exposed by day on a lamppost, wall, or porch ceiling. The vast bulk of them, however, are hiding in their natural habitats and escape human detection, even when we go looking for them.

We might want to find them out of curiosity. Birds search for them as something to eat. The only rational explanation for the striking resemblance of so many moth species to the backgrounds where they live is camouflage.

In appropriate surroundings, moths are either hard to see (cryptic) or are mistaken for something else (mimetic). The distinction between crypsis and mimesis is not always clear, but both forms of disguise work whenever a predator using vision to locate a meal fails to recognize a moth as food.

Protective coloration alone isn't enough to escape detection. Because movement will betray an otherwise well-concealed moth, it is not at all surprising that moths remain dead still when hiding. Nonetheless, hungry birds are determined hunters and patrol their habitats in search of food. Sudden death is the likely price a moth pays when it's discovered, but those not grabbed onto firmly by a predator sometimes escape through tricky backup maneuvers. Moths in the genus *Catocala* are famous for flashing showy hindwings that are brightly colored, compared with their darker, cryptically patterned forewings, under which the hindwings are normally hidden from view while the moth is at rest. Almost any physical disturbance triggers these underwing moths to take instant flight. No warm-up required! The putative protective value of this startling display during flight is that predators are confused into searching for a color pattern that disappears the instant the moth has landed and its underwings are again tucked away. Distraction tactics appear to be the backup behavior for *Catocala* when crypsis fails. Other groups of moths use different actions, such as displaying giant "eyespots" or extending their bodies in colorful bluffs. The degree to which these protect against predators is much discussed among experts, and amateur collectors enjoy demonstrating these various behaviors to amazed onlookers.

By comparison, peppered moths have very little to offer in the way of backup behaviors. In fact, they are remarkably resistant to disturbances once they have clamped down for the day. Just after dawn, however, they are still pretty skittish and will fly away if disturbed. This usually involves warming up, by flapping their wings in place before takeoff, so it is not likely to be a successful escape tactic from a cold start. At rest in a comfortable spot (meaning out of direct sunlight and on a suitably textured surface), peppered moths are harder to wake up than a teenager on a school morning. You can poke and prod them and they stay put, convinced (if I may lapse into anthropomorphism) that they are still invisible. Maybe their backup strategy is to deny that they have been discovered, stubbornly insisting to a would-be predator, "I'm not really here, so go away!"

If one is determined to rough them up, they will drop to the ground and play dead, sometimes squirting out a brown liquid, which is metabolic waste that has been stored since eclosion (emergence from the pupal to the adult stage of the life cycle). I put my tongue on some fresh brown excreta to see if it tasted bad or bitter and might act as a deterrent, but nothing registered for me. I also know from numerous observations that birds eagerly eat peppered moths, brown juice and all. For this species, hiding is the best defense against predators, and the backup ruse is to remain immobile, even when discovered. The payoff may be that a chance encounter by a predator does not necessarily result in recognition if the moth stays still and the unsuspecting predator moves on. "Phew, that was a close one," the moth might sigh to itself. It probably doesn't "think" in ways we understand, but a disturbed moth is an alert moth, and it will bolt if the harassment soon returns. Otherwise it moves as little as necessary until nightfall. Then it's party time!

Animal coloration is a vast topic and can involve all sorts of things, including thermal regulation, diet, sexual selection, aposematism (warning coloration), mimicry, and camouflage. Camouflage evolves from predator-prey interactions. By definition, it means one thing is hiding from something else. Camouflaged predators are protected against early detection by their prey. Camouflaged prey are concealed from their predators. These are not mutually exclusive conditions in animal species that play both ecological roles. Clearly, peppered moths are not predators. They don't even feed as adults. Yet they are prey. Thus the conclusion that predators have driven the evolution of their camouflage is inescapable. What predators, then, are the most likely candidates?

Bats catch flying peppered moths at night, apparently in large numbers, but could such nocturnal mortality exert selection pressure on the moths' daytime resting habits and wing-color patterns? Bats generally find their prey by echolocation, not vision, so the idea that bats were driving the evolution of melanism in peppered moths lacked logical support. Nevertheless, the bat question was resurrected from time to time, until Michael Majerus finally buried it with data. He released equal numbers of melanic and pale peppered moths at several locations where bats were feeding. The total numbers of melanics and pales eaten by the bats over three seasons (2003–2005) were 208 and 211, respectively, demonstrating convincingly that bats do not discriminate between the peppered moth phenotypes.

Mice, and even ants, can also nab resting peppered moths, day or night, but it is unlikely that camouflage would protect the moths from this source of mortality. The protective coloration of peppered moths—and this probably holds true for cryptic moths in general—has evolved by natural selection, in response to the pressures exerted by predators that use vision to find prey. Namely, birds. What else could have done it? (The extensive experimental evidence supporting this conclusion will be examined in a later chapter.)

Many moth species ordinarily rest on natural backgrounds that are appropriate to their appearance. The question is, do they settle randomly on whatever they happen to encounter (with the mismatches getting eaten), or do moths actively select hiding places? Several independent studies have convincingly demonstrated that the latter occurs. While various factors, such as odors and textures, may serve as cues to guide a moth to its resting site, surface reflectance has been firmly established as at least one of the components involved. Direct tests of rest-site selection were inspired by Kettlewell's famous barrel experiments, published in the journal *Nature* in 1955. Studies of the behavior of captive moths kept in suitable containers, or holding pens, are called "barrel experiments," because the original enclosure was a large cider barrel, lined inside with contrasting colors. Kettlewell's experiments were originally designed to compare the behavior of polymorphic peppered moths (to be discussed in detail in the next chapter), but the basic design of his experiments has been used to examine the behavior of several dozen monomorphic species representing the two largest moth families, the Noctuidae and the Geometridae. In 1966, American lepidopterist T. D. Sargent—and, later (in 1974), the British team of Boardman, Askew, and Cook—showed that different species of captive moths, when presented with backgrounds of different surface reflectance (light vs. dark) chose the settings most closely matching the reflectance of their forewings. The generalization, however, is based on strong statistical correlations, not perfect relationships. Mismatches are common. Critics have argued correctly that captive moths are surely not behaving normally, and that other factors besides surface reflectance are involved in rest-site selection. Kettlewell himself was the first to point out the shortcomings of his own experiments. But in defense of barrel experiments and what they can tell us, moths are clearly not indifferent to the backgrounds offered to them in captivity, however artificial these may be. While reflectance matching between a rest-site substrate and body color is broadly true across moth species, the exceptions are instructive. The Boardman et al. study demonstrated that moths mimicking the color of bird droppings preferred dark backgrounds. One needn't actually rest on poop to be mistaken for poop. There is a lesson here.

5 | The Rest-Site Selection Controversy

Newspaper headlines periodically announce "EVEN SCIENTISTS DISAGREE" about the meaning of some recent finding, as if a lack of concurrence among scientists is, in itself, unusual. Scientists disagree as a matter of routine. It's an important part of the process that forces them to make progress, which is familiar ground in essays about how science works. In a nutshell, just about anything anyone claims to be true will be challenged, and whatever evidence might be provided will be painstakingly dissected. Eventually.

Some controversies in science last for decades, or even longer. But without new, relevant data, old controversies grow stale and are retired, not resolved. Clever arguments alone cannot settle controversies. Theoretical models lasting too long without empirical support get ignored. The only way to decide an issue—for now, if not once and for all—is to conduct the critical experiment or perform the necessary observations.

That was my plan for using the peppered moths at Mountain Lake. I intended to carry out the essential research that would resolve the back-burner controversy between Kettlewell and Sargent regarding the *mechanism* by which moths choose their daytime hiding places. Given that moths nonrandomly select surfaces with different reflectances, how do they do it?

In his original barrel experiment, Kettlewell put six peppered moths of the same sex, but equally divided by phenotype (three typicals [pale] and three *carbonaria* [melanic]), into an enclosure made from a wooden barrel. The sides were lined with alternate strips of black and white heavy paper having the same texture, and the top was covered with a glass plate, where the moths could not easily gain a foothold, thus forcing them to take up resting positions on the sides of the barrel. The apparatus was placed in shade, under trees, and kept there overnight. The following morning the positions of the moths resting on either the black or the white strips were recorded. Moths overlapping the alternate surfaces were not scored as having made a "choice." Over the course of of the experiment, using fresh moths each night, Kettlewell's sum-mary data for 118 choices are as follows: *carbonaria* selected black 38 times and white 21 times; whereas *typicals* were on black 20 times and white 39 times—nearly mirror-image results.

Kettlewell's data indicate that background selections by different moth phenotypes within the same species are not independent of their body colors. Again, this is a statistical inference, not an absolute. While mismatches were common, they were significantly in the minority. Thus the null hypothesis—that no behavioral differences exist between the phenotypes—is rejected by contingency chi-square ($P = 0.001$), a test used to calculate the probability that the observed difference in background selections made by the two moth phenotypes is due to chance alone. What, then, might produce the bias shown by dark moths for dark backgrounds, and the preference pale moths have for light backgrounds?

Kettlewell suggested that *morph-specific* background selection is the end result of a behavior he called "contrast/conflict." According to this idea, moths compare their own body parts with the surface on which they have landed. In Kettlewell's (1973) own words:

> At early daybreak . . . a tree of the correct species [Kettlewell suggested this was determined by scent] is chosen at random and on alighting the moth runs over the trunk. Here it shifts its position maybe by several feet, so as to judge the best of the local advantages offered. This it does by vision. Having discovered its own optimum it now carries out a particular behavior pattern which I believe has previously not been described. The insect turns on its own axis, maybe through the whole 360°, and during this act it from time to time "clamps" its wings flush with the trunk. The final movement is a side-to-side wiggle when it succeeds in getting its thorax and abdomen aligned into a groove. I have made these observations from several hundred moths released at earliest daylight and subsequently followed to their destination. I think that during this sophisticated act several different stimuli are being received and their correct answers determined. First, wind direction; no Lepidoptera chooses the windblown side of a trunk. Secondly, and more important, during the "clamping" movements the insect is receiving light stimuli from its background which it can compare with the colour and pattern of its own circumocular tufts. If the two are out

> of phase . . . a state of "contrast/conflict" will occur . . . [and] the insect will then move to another position or even take flight again. When once the final positioning for the day is decided, considerable disturbance may be necessary in order to stimulate the flight reflex. This is certainly so in *Biston betularia* where the usual behavior is to feign death in a period of catalepsy. (pp. 71–72)

Although Kettlewell did not use the same terminology, his proposed contrast/conflict mechanism of background selection would qualify as simple kinesis, stimulating a moth to move about until the contrast between itself and the reflectance from its surroundings falls below an internal threshold. As barrel-inspired experiments with monomorphic species have established, moths are sensitive to the reflectance from their surroundings when choosing resting sites, but no self-inspection mechanism has been proposed in those instances. Nor would self-inspection seem necessary, as all of the individuals in monomorphic species essentially look alike. What is new here is the proposal that the different phenotypes in a polymorphic population must ask themselves, figuratively speaking, "Which particular color am I?" They do this by comparing themselves with their backgrounds, to see if they correspond. If so, then all is well and good, but if not, the moth keeps moving until a matching background is encountered.

Envisioning background selection by moths as a form of kinetic behavior was a brilliant idea. Moreover, Kettlewell suggested that such behavior could facilitate the rapid rise and spread of melanism. A novel phenotype, introduced into a population either by mutation or migration, would automatically do the right thing when it came to finding appropriate places to hide (unless all potential hiding places were of uniform reflectance). The genetic newcomer would be off to a running start, as it were. If, instead, a novel phenotype had a predetermined (hard-wired) behavior, programming it to do the same thing the common phenotypes did, it would be conspicuous and most likely picked off by a predator before living long enough to reproduce and pass along copies of its special genes. With contrast/conflict, however, the very same gene producing the melanic phenotype also influences the moth's behavior in predictable ways. Geneticists call genes that influence more than one phenotype (e.g., both body color *and* behavior) “pleiotropic.” This particular pleiotropism would be highly adaptive: darks moths nonrandomly selecting dark places to hide, while pale moths do just the opposite. Nice!

But does it work that way? Kettlewell attempted to test this contrast/conflict idea by hoping to fool the moths about their true colors. Moths and butterflies are members of an order of insects called Lepidoptera, meaning “scaled wings,” although scales cover pretty much most of their bodies. “Circumocular tufts” are long, rather wooly-appearing scales surrounding the eyes. Kettlewell wanted to reciprocally graft these tufts between pale and melanic peppered moths, so that what the recipient moths would see were scales supplied by donors of different colors. The experiment failed for lack of a suitable technique, as Kettlewell acknowledged. If you try to perform the operation yourself, you will easily understand the problem. It’s not hard to remove the scales from a moth, but it’s no fun trying to glue them back on, especially for the moth.

That’s how things stood on this question of contrast/conflict for over a decade, until Sargent decided to have a go. Instead of transferring scales between moths of different colors, he decided to use paint to coat intact scales. Good idea! He painted some white moths black and black moths white, as well as coating others in their own shades, as controls. To keep the moths still while he performed this operation, he knocked them out briefly with cyanide. Some of them remained unpainted, to assess the effect of the anesthesia on their behavior. Still others were untreated (no paint and no anesthesia), again to serve as controls. Sargent put his moths into his own version of a test barrel, to see which background (black or white) they might choose. The bottom line is that his painting treatment did not make any difference in the moths’ preferences. Naturally dark moths, whether painted white or not, still selected the black background, and naturally white moths, whether painted black or not, still chose the white one. It was clear from his experiments that these particular moths were not using self-inspection to choose backgrounds—or, at least, they weren’t fooled by the paint. The behavior of the moths in Sargent’s experiment appeared to be genetically fixed.

If *Biston betularia* was indeed showing morph-specific background selection behavior, Sargent argued that such behavior was probably hard wired by genes that were nonrandomly associated with the genes that determined body color. Separate genes that do different things but are expressed together as adaptive complexes are referred to as "coadapted." Unlike pleiotropy (with one gene influencing two or more traits), which arises simultaneously with a new mutation, the assembly of coadapted complexes of two or more separate genes would probably require generations to become established as commonplace in a population. Sargent supported his argument by comparing a stable, polymorphic species, *Catocala ultronia*, for which he reported morph-specific background selections, with *Cosymbia pendularia*, in which dark phenotypes occur only sporadically and do not differ in their background choices from the typical (pale) phenotype. Sargent's discussion about coadaptation explaining morph-specific behavior was brief and circumstantial, but his evidence that the particular background selection by the moths he painted argued strongly against a contrast/conflict mechanism operating in the species he tested.

It might seem that contrast/conflict was by then a dead concept, ready for burial, and should have been forgotten, along with all the other good ideas in science that have failed to gather empirical support. But no! Kettlewell was quick to spot the fundamental flaw that disqualified Sargent's experiments from having the final word. Instead of using *Biston betularia* to test contrast/conflict, Sargent painted the scales on two separate monomorphic species of moths, *Campaea perlata* (a pale geometrid), and *Catocala antinympha* (a dark noctuid). That he used different species of moths, with one species from a different taxonomic family from *Biston*, is not the relevant point. The key issue Kettlewell seized upon was that Sargent did not use polymorphic moths, which showed morph-specific behaviors, to test contrast/conflict.

Why was this so important? By definition, all members of a monomorphic population qualitatively have the same phenotype. Therefore, they would be comparably camouflaged by the same backgrounds. Long periods of natural selection favoring similar responses to similar backgrounds could genetically fix that behavior. Conversely, all members of a polymorphic population do not look alike. Should they also display morph-specific responses to different backgrounds, self-inspection may or may not be part of the process of background discrimination. Kettlewell's point was that Sargent's failure to find contrast/conflict in monomorphic populations with recent evolutionary histories of consistent selection for uniform behaviors did not preclude a role for contrast/conflict in polymorphic populations that are correspondingly polymorphic in their behaviors. The question of whether the morph-specific responses resulted from pleiotropy (one gene with manifold effects) or coadaptation (different genes for pigmentation and for behavior) remained unanswered.

Science journalist Judith Hooper, in her 2002 popular-literature book, *Of Moths and Men*, spawned confusion over this controversy. She reversed the arguments by attributing the coadaptation hypothesis to Kettlewell (p. 272), and she accused Kettlewell (and me) of falsely claiming Sargent never worked with polymorphic moths in background-selection experiments (p. 286). Kettlewell's claim was context specific, and Hooper clearly missed that point. While Sargent worked with polymorphic moth species, he failed to test these polymorphic species for the contrast/conflict mechanism. He did not, for example, publish (to my knowledge) experiments where he painted the scales of pale and dark *Catocala ultronia*, the species in which he had reported morph-specific behavior. Why not? That would have been exactly the experiment Kettlewell would have demanded of him. It's still waiting to be done. There would have been no point in painting *Cosymbia pendulinaria* or *Phigalia titea*, as these two polymorphic moths were behaviorally monomorphic, and no one is faulting Sargent for not painting them. Unfortunately, Hooper did not master the rudiments of this controversy, but she attempted to explain it anyway.

Both Kettlewell and Sargent published their papers on background selection in the most visible journals on the planet, *Nature* and *Science*. This particular controversy was not buried in the literature of specialists. It was out there, man! Still, the issue remained unresolved. But that's not unusual in science. Was this just another game not worth the candle?

One day, many years ago, while lecturing about industrial melanism, I mentioned this controversy to my class. Just about any example scrutinized beyond standard accounts is replete with unresolved issues, and textbooks would be unwieldy if these were all explored. Nonetheless, a couple of enterprising students interrupted me with a quote from E. B. Ford's famous treatise, *Ecological Genetics* (which was by then in its third edition, published in 1971): "Moths tend to take up positions in which their colouring matches their background. . . . It seems that this is done by recognizing the contrast between the colour of the circumocular tuft of scales and that of the bark or lichen on which they rest" (p. 301).

Ford properly cited Kettlewell's widely read *Nature* paper from 1955, but he made no mention at all of Sargent's doubts about contrast/conflict, which appeared in *Science* in 1968. Books are always out of date even before they are published, because their production is a lengthy process, but Ford still did not revise this statement in the fourth (and final) edition of his great book, using the exact same words again in 1975 (p. 325). Even if Sargent's work utterly escaped Ford's notice, he most certainly knew that Kettlewell himself had never actually tested contrast/conflict, for lack of a viable technique. Thus the idea of contrast/conflict remained just that: an untested, unconfirmed *idea*. Yet here, in the definitive book for the field by its preeminent scholar of that period, who also happened to be Kettlewell's mentor, Ford's choice of words hinted that this mere suggestion had at least some supporting evidence, elevating it beyond the idea stage to the seems-that-this-is-done-by stage. It was entirely appropriate for Ford to entertain thoughts of how contrast/conflict *might* work, but he had a responsibility to his readers to remind them that it was pure speculation.

My students were disappointed that world-renowned scientists could so uncritically accept and promote untested hypotheses, and they learned a valuable lesson: just-so stories don't come only from amateur naturalists. This one was launched by the pros—the best ones in the business.

By the time I got to Mountain Lake, eager to clean up this untidy brouhaha, 15 years had passed since it held center stage. That there ever *was* a controversy had been all but forgotten. Most textbook accounts about industrial melanism either didn't mention morph-specific background selection, or, if they did, little (if any) attention was paid to the mechanism. This lack of awareness about the controversy I intended to resolve in no way dampened my enthusiasm for the work that lay ahead. I could think about little else that summer of 1983.

6 | A Feeling for the Organism

Bernard Kettlewell died four years before I had my first peppered moth in hand and wondered what to do with it. I would have liked to ask his advice.

There is speculation about how Kettlewell's end came. As is true for most people who've led full lives, he experienced sadness and loss, and toward the end of his life he had suffered severe physical discomfort. I've heard the rumor of suicide from more than one source, but I don't know if it's true. Perhaps I should have asked Cyril Clarke about it after I got to know him better in later years, but I didn't know how to broach the subject. I felt it would be prying, and I wasn't entitled to that privilege. They were close friends, and Cyril wrote Kettlewell's obituary, making no mention of suicide. I do know, however, that Judith Hooper (Kettlewell's notorious biographer) was unaware of that bit of hearsay until I naively mentioned it to her during her telephone interview with me. She paused for a moment as the information registered, then said, "That's interesting. Hmm, now that's *very* interesting." Worried about how she might exploit my faux pas, I immediately cautioned that she shouldn't read too much into the talk of suicide, because (1) it might not be true, and (2) his health was declining. But by then the damage was done. Hooper seized on this scuttlebutt and later portrayed Kettlewell as a tormented soul who put himself out of his own misery. High drama, but is it accurate?

I bring up Kettlewell's death as a bump in the road to making scientific progress. Instead of our being able to visit him and ask, "How'd you do this?" or "Why'd you do that?" we are forced to guess the answers and frame questions about what to do next without benefit of his vast experience. We had lost the opportunity to learn from him, beyond what he'd published.

Even had he left detailed field notes, they still would have been a poor substitute for personal, hands-on experience. Just as serious musicians or artists study with masters, so, too, do scientists who are seeking to enter a new field or learn a new technique. Aspiring scholars methodically visit those who already know what they hope to learn. There's no sense in reinventing the wheel, or, in this case, the barrel.

My first attempt at repeating Kettlewell's barrel experiment was to start with a much bigger and softer-sided barrel. Among Kettlewell's (1973) complaints about Sargent's experiments was the small size of the apparatus the latter had constructed as his test pen for background selection, as well as the number of moths he had put into it all at once: "Without having had the advantage of seeing [Sargent's] techniques, and therefore with some reservations I make the following criticisms: the containers were much too small (approx. 15 in. square by 35 in. high), . . . the insects were overcrowded (up to 10)" (p. 70). Indeed, Kettlewell candidly expressed dissatisfaction with his own practice of cramming too many moths into a confined space: "A large cider barrel (height 40 in., maximum diameter 28 in.). . . . I came to the conclusion that even six individuals were too many to use at any one time because of disturbance effects" (p. 69).

To remedy that problem, I thought big. I purchased a screened dining canopy, measuring about 12 feet wide on each side and tall enough to stand in. My plan was to hang black and white cloth "flags" all around the inside of this screen house, to give the moths lots of surfaces on which to take up rest sites without interfering with each other. I envisioned how the tests might run: once the moths settled onto the surfaces, I and my field assistants (my two daughters) would count how many moths of each phenotype were on the black or white stripes. After scoring, then we'd shake the flags sufficiently to get the moths flying again, to start another run. That way, I hoped, we'd get lots of data in just a few days. Sounded easy.

Two things were wrong with this experimental design, and both were the result of my lacking what the Nobel Prize–winning geneticist Barbara McClintock had for maize (corn plants), "a feeling for the organism." One doesn't acquire such intimacy via *Gendankenexperiment*. Until that summer, my knowledge of animal behavior was lab bound. I was guilty of learning about biology by observing fruit flies "cavorting in glass bottles," as Stephen Jay Gould once put it. Within minutes of our first trial run, my daughters and I immediately recognized that our procedures were flawed, and we were doomed to fail if we continued to pursue that route. "Poor Dad," they must have been thinking but mercifully didn't say. All of the moths, every last one of them, landed and stayed on the screen walls of the tent, and none—not a single one of them—took up a resting position on either the black or white flags. Nada!

"Don't let this get you down," I reassured my daughters. "We can fix this problem. Easy!" They glanced silently at each other as I went on to explain that we'd just cover over the screen walls—top, bottom, sides—with black and white cloth, so when the moths landed on the outer walls of the pen, no matter where, they'd be confronted with black and white surfaces. No other options would be available. Period! Then we could count how many of what kind stayed put on either the black or the white. By this time my daughters had regained their enthusiasm for science, although they told me they could hear the dinner bell ringing, meaning lunch was ready at the dining hall.

Meals at Mountain Lake were not to be missed. None of them, if you could help it. The kids—whatever their ages, from toddlers through graduate students—liked to socialize. Who doesn't? And the resident faculty, visiting scientists from other universities, and postdocs enjoyed bouncing ideas off one another. It's a familiar scene, played out across the globe wherever scientists gather. Science really is a very social activity, at least in some aspects, and it is a lot of fun. More importantly, we learn from one another. That happened at Mountain Lake each day, at every meal. I was just as interested in the dinner bell as my daughters were. We all gained weight, too, except I wasn't growing any taller.

The second flaw in my experimental design was subtler, but easily as important: moths disturbed during the daylight hours do not behave normally. They will land on just about the first thing they come to and quickly clamp down, quite unlike the behavior Kettlewell described them performing at dawn. When moths are rousted out of a "deep sleep" in full light, their instinct to not move during the day overwhelms their predawn tendency for habitat selection. It soon became clear that we would get only one background choice per day for our captive moths at Mountain Lake. We'd have to put the moths into the redesigned test pen before nightfall, then score their positions in the pen the next morning. This meant the experiment was going to take a lot longer than I'd hoped. It required several more years, in fact. As Dan Grosch, one of my old professors, dryly consoled whenever someone's experiment failed, "Nature reveals her secrets reluctantly."

7 | Elizabethan Moths

My test pen (or barrel) went through its own evolution as I continued to tinker with it while acquiring a feeling for the organism through intimate experience. At Mountain Lake that first summer, I had used a prototype test pen, based on a scaled-down version of the screen house, with black and white vertical stripes sewn into the sides. The exterior was covered with cardboard panels, so incoming light could be directed where I chose. It was essential for the light to be reflected from, rather than pass through, the surfaces on which the moths took up resting positions. Each morning I'd record the locations and phenotypes of the moths in the pen.

What I needed to learn next was how to fool a moth about its inherited color. I started by attempting Sargent's approach: painting the scales surrounding their eyes. This proved to be a lot harder that it sounds.

Moths don't like being painted. They refuse to hold still. Sargent used cyanide for an anesthetic. Some collectors employ it to kill moths, so one must stop short of that result in order to paint them and hope they survive the treatment well enough to make background selections inside a barrel. It was pretty rough.

There are better anesthetics for insects. As a *Drosophila* geneticist, I've employed both ether and carbon dioxide (CO_2) routinely for years. I prefer CO_2 because, unlike ether, it doesn't explode, and I've used it successfully on wasps, too. Why not try it on moths?

CO_2 is readily available in pressurized cylinders that can be purchased from local suppliers, and I had some delivered to my lab at William & Mary on a regular basis. It was easier to obtain on government contracts than ballpoint pens that actually worked. But now I was at a field station a long way from home. Most of the scientists who went there knew beforehand what they would need for their work and carried the necessary equipment with them. Although I brought along a lot of gear that I'd guessed I might need for the project, my requirements changed as I started to learn more. I should have made prior arrangements to have CO_2 delivered, but I naively assumed there would be a convenient nearby source. There wasn't, and I wanted some immediately. I had live moths waiting to be painted. I couldn't risk killing them using harsh anesthetics. Luckily, I was at Mountain Lake, where whatever you needed could be dug up by asking for help.

Jim Murray's secretary, Mary Ann Angleberger, was one of the people to ask. Another was her husband Wayne, who oversaw the sparse stockroom. The couple worked as high school teachers during the academic year, but they liked to help out at Mountain Lake and returned each summer for many years. Mary Ann had earned a master's degree from the University of Virginia and was a prize-winning naturalist. The station held annual contests, and she was an acknowledged champion. She also was observant enough to recognize that my eight-year-old daughter needed something to occupy her time that summer, so Mary Ann got her involved in crafts with other Mountain Lake "brats" who were at loose ends while their parents worked in the woods.

Wayne was full of lore about the place and knew just where to send you if you needed something out of the ordinary, such as pressurized CO_2. So down the mountain I drove to Pembroke, to a small factory that used CO_2 in its welding operations, and inquired, "Do you have any CO_2 to spare that I might buy from you?" They didn't even ask me what I wanted it for. Just knowing I was from the biological station was good enough for them. They gave it to me for free.

"Just return the tank when you've finished," was all they said. Nothing to sign. Just a shake of the hand. I liked doing business in Pembroke.

Back in my lab at the station, I zonked out one peppered moth at a time and painstakingly painted its scales. They survived the CO_2 anesthetic without any discernible problem, as I strongly suspected would happen from my previous experience with other insects, but I never found the right paint for the job.

The painted moths looked pretty good when I put them into the test pen before sunset. I had a standard method of starting a "run" by placing them on a tray and setting it on the floor of the pen. Untreated moths would leave the ground-level tray during the night and bat around the sides and top of the enclosure until they finally settled on a place to rest before sunrise. This level of activity depended on the prevailing conditions, but it was clearly the case that, by morning, healthy moths would seldom be where they were initially placed. They would nearly always be clamped onto either the top or the sides very near the top. Moths on the floor were usually debilitated in some way and often near death. This is close to a qualitative generalization, like what happens in a horse race. Horses almost always leave the starting gate, unless something is very wrong. The same is true of moths not leaving their "starting gate," the ground-level tray in the test pan.

The grand bulk of my painted moths stayed on the floor. A few of them moved onto the black and white panels above, but not many. Unpainted moths, even the controls that had been anesthetized with CO_2, were all over the place, but rarely on the floor. The conclusion I drew from this was that painting does affect moth behavior. It makes them sick. Clearly I was using the wrong paint.

I tried different kinds. A lot of different paints: water-soluble acrylics, nontoxic this and that, quick-drying dopes, lacquers, pigmented powders, talcum, corn starch, anything I could find in craft stores all over Pembroke, Blacksburg, and Roanoke. The best results I got were that some moths moved away from the floor of the test pen, but by morning these had no scales, painted or otherwise, surrounding their eyes. The paints had stiffened the scales to the point they had fallen off as the moths batted around inside the test pen. The color they would then see would be that of their bare "skin," the beige of exposed chitin cuticle that was entirely denuded of scales by morning. At this point I was suspicious that the photographs Sargent published of his painted moths were taken before, but not after, they had spent the previous night banging around inside a container. What did they look like the next morning?

I tried unsuccessfully to locate a source for the paint Sargent identified in the paper, published some 15 years previously, when he tested Kettlewell's contrast/conflict hypothesis. Craft-store people either said they didn't recognize the brand or suggested it might no longer be available. How unique could it be? In despair, as I was running out of time and moths during my brief stay at Mountain Lake, I went to the office to see if Mary Ann might be able to connect me with potential suppliers elsewhere.

When I entered the room, I noticed that she was busily cutting holes in notebook paper, using a hand-held punch. The punched-out disks that are normally discarded caught my eye. The bits littering her desk happened to be white. But they could've been black, too, had she chosen to punch holes in black paper. It hit me all at once, I imagine something like Nobel Prize–winning microbiologist Joshua Lederberg's idea for replica plating (pressing a piece of sterile cloth onto one Petri plate of bacterial colonies and using it to transfer colonies from the original plate to other plates) must've hit him. Eureka! Admittedly my questions, by any standard of comparison, were minuscule compared with Lederberg's ingenious method for determining the independence of bacterial mutations and their exposure to selective agents, but suddenly I saw a way around the problem of using paint to fool moths about their inherited colors. I'd put colored paper collars around their necks to prevent self-inspection. As with many of my minor equipment needs, Mary Ann again came to the rescue. I think I still have her paper punch.

The paper punch produced nearly perfect circles, 6.5 mm in diameter, that I used to construct collars about twice the width of the average peppered moth's head (3.2 mm, as measured from the outer edges of the eyes). When resting normally, untreated moths (those without paper collars) have an unobstructed view of a mass of "wooly" scales on the proximal joints of the forelegs and points aft. With a paper collar in place, this view to the rear is blocked, although the moth can still see the substrate in front of its head and the underside of its antennae. By itself, a moth's head is relatively devoid of scales, except for a sparse distribution fringing the eyes, and those few not hidden by the collar are easily touched up with tiny dabs of paint, without obvious ill effects. If Kettlewell's self-inspection hypothesis was correct, a collared moth would compare the substrate it's standing on or walking over with the color of its paper collar, rather than the expanse of body scales (its true color) normally in its field of view.

Fitting moths with paper collars turned out to be remarkably easy, once I figured out how to accomplish it. Moths have very short, skinny necks. Though threadlike in thickness (compared with the head and thorax it connects), the neck is strong enough to hold these essential body parts together, even after being fitted with a paper collar. In terms of vigor and longevity, the moths seemed totally unaffected by the collars. Plus, they looked rather rakish in their Elizabethan garb, evoking images of Sir Walter Raleigh or Shakespearean moths.

The technique required inserting a paper disk between the moth's head and thorax. I couldn't think of any way to pull a paper collar over a moth's head and antennae without making a neck hole in the collar wider than the head. That wouldn't do. The solution was to construct a completed collar by using two paper disks, each with a section cut out, like a pie with a single slice removed. Each half resembled a Pac-Man with its mouth open. The openings allowed two paper disks to be slipped between the head and thorax from opposite directions and overlap, forming a complete disk that encircled the head. The two pieces were then pasted together with a smidgen of Elmer's Glue.

Into the test pen went black moths wearing white collars, pale (typical) moths wearing black collars, black moths with black collars, pale moths with white collars, and moths of both phenotypes with no collars, to see which ones chose which backgrounds (black or white) as their rest sites. At last, I hoped, my experiment was finally underway.

It's a funny thing about experiments. To paraphrase the Rolling Stones, you don't always get what you want.

8 | Nonrandom Rest-Site Selection in Captivity

What I wanted but didn't get for my experiments at Mountain Lake were enough moths. The trapping yields would have easily pleased general moth collectors looking for gorgeous specimens from a wide variety of species, but the total number of melanic peppered moths was far too low for my focused purposes. The percentage of melanics in our sample was actually above average from previous years, but this still wasn't enough for me. In all, we had collected 148 peppered moths, of which seven (4.7%) were melanic, the highest percentage recorded at that location since Dave West started trapping there in 1968. But, come on! Seven melanics could hardly be assigned to three categories (those wearing white collars, or black collars, or no collars) to assess the effect of collar color on rest-site selection. The best I could hope for was to compare untreated melanics with untreated typicals, to determine if significant differences between the phenotypes might exist in their background choices. The self-inspection hypothesis would, by necessity from the limited number of appropriate moth subjects, be restricted to comparing the behavior of pale phenotypes wearing collars of different colors.

Although Dave regularly joined me at the moth trap at dawn, he didn't always have the luxury of time. He was teaching a course at the station, whereas I had only one thing on my mind. After he showed me my first peppered moth (the pale phenotype), and quite a number of pale specimens after that, I was on my own. I was perfectly confident that I wouldn't miss any in the trap. They were remarkably easy to distinguish from all the other moth species in the vicinity, except maybe (for rank beginners) oak beauties (*Nacophora quernaria*), another geometrid strikingly polymorphic for melanism. After days went by with still no melanic peppered moths, I showed up at breakfast one cold, rainy morning and bluntly asked Dave, "How will I recognize with certainty a melanic when I see one?"

He looked up from his tray of steaming chow and said, "They're black."

More days passed with no melanics, but by then I had collected a few more healthy typicals (pale moths) to test in my background-selection pen. I had been running a few trials all along, but until then I hadn't settled on a suitable test-pen design to conduct the background-choice experiments. When I saw Dave at breakfast, I told him my first "real" results. I'd put seven uncollared moths into the pen the night before, and all of them were on black stripes in the morning. "Not random," Dave said.

"Right," I said. "Point five to the seventh is less than 1%." Still, I worried, it could just be a fluke.

But the pattern continued. Although some typicals were on white stripes in the morning, a statistical story was emerging. American peppered moths, in this case the pale phenotypes, showed a bias toward resting on black stripes, as opposed to white stripes. It wasn't random. Actually, the American typical phenotypes aren't really very pale, compared with their British counterparts. Kettlewell had pointed out years before that American typicals more closely resembled the British intermediates called *insularia*.

I decided this wasn't altogether bad. Our darkish pale typicals preferred to hide on black as opposed to white because they didn't really look white. Indeed, they were far more conspicuous on white stripes than on black stripes. I should have used a grayer tint for the light-colored stripes and paid more attention to the reflectance measurements taken by Sargent. Nevertheless, the moths showed a clear preference for one of the two background colors equally available to them. Whatever the shortcomings of barrel experiments, and however artificial the environment in which these moths were held captive, they were definitely not settling on the stripes at random. Something was going on.

When given a choice between black and white, American peppered moths showed a strong bias for black. Was this behavior fixed by genes, as Sargent suggested? Or did it result from self-inspection, as Kettlewell proposed? Perhaps I didn't really need melanic specimens to answer the question if I collared the black stripe–preferring typical American peppered moths in white versus black collars and statistically compared their background selections. I was heartened to be still in the game. And, as a bonus I'd use whatever melanics happened into the trap, to see if their preference for black might be even stronger. Exciting times lay ahead, whatever the outcome.

9 | Life at Mountain Lake

By the end of that first summer at Mountain Lake, I had answered the simple question I went there to ask: Do peppered moths select resting sites by the contrast/conflict mechanism, as suggested by Kettlewell? The answer: no. Frankly, I had wanted a different answer. Kettlewell's idea was so good, I'd hoped it was true. But, alas, my data did not support the hypothesis that self-inspection was part of how a moth finds its hiding place. This was not a vindication of Sargent's work, however, as his experiments didn't really address this question.

Still, I was uneasy about my results and how I went about getting them. I had little confidence in my data at that point. Part of the problem was simply sheer numbers. I had caught too few peppered moths to do the experiments properly, having spent so many specimens on pilot studies in attempts to design experiments that would yield unambiguous answers. The upside is that I gained valuable experience that would serve me well over the years, as I continued to acquire a feeling for the organism. I was becoming acquainted with the star of this drama, *Biston betularia*, eyeball to ommatidium. The aphorism that only fools graduate from the school of experience is believed by those with too little experience. There is no question that we learn from the accomplishments and failures of others. Only a fool would argue otherwise. But there is no substitute for personal know-how, the aphorism believed by those with hard-won diplomas from the school of experience.

For me, the lessons at this school were quite pleasant, done in agreeable surroundings, in company with congenial people. Each morning I arose at first light to beat the birds to the moth trap. Sometimes curious early risers would stop by to see the assortment of moth species in the trap and speculate or inquire about their various color patterns, or ask for their identities. I was usually stumped by their questions, as I was still new to moth hunting. I know a bit more now, but I'm no less amazed at the great diversity and beauty among moths. Most of the time it was too early for all but a few hearty souls to venture out on chilly, misty mornings. Yet there were a few adventurous spirits at the station who would swim in the cold pond (Riopel Dam) before breakfast. These included the teenage daughters of Jim Murray, Dave West, and yours truly.

After dealing with the trap, I'd immediately go to my screen house in the woods to score the moths that I had put into the background test pens the previous evening. Then it was time for breakfast, after which I went to the lab to put collars on new moths for the next night's run. This usually didn't take all morning, and some of that time was spent trying different techniques, fiddling with the pens, making repairs, or reading the stack of research papers I'd brought along. I also had plenty of time to chat with people in other labs about what they were working on, and some gave me guided tours of their research setups. My most significant interaction was with Jim Murray's new graduate student from Japan, Takahiro Asami, or Hiro, as we called him. Hiro had just arrived at Mountain Lake that year to begin a long-term study on snail populations for his PhD dissertation. (I will have much more to say about him in later chapters, as he and I eventually became brief collaborators in moth work and have remained good friends to this day.)

Afternoons were much like the mornings, except I'd take occasional breaks from work to spend time with my daughters. Sometimes we'd go for long nature hikes in the woods, usually accompanied by lots of questions, such as "What kind of fern is that, Dad?" After I failed to answer most of the queries they put to me about plant identities, my older daughter, Megan, finally asked, in exasperation, "What kind of biologist are you, Dad?" Sometimes we'd go for a swim, either in Riopel Dam at the station, but preferably in the real Mountain Lake, which (in 1983) was crystal clear and fantastically refreshing. I couldn't linger in the frigid water for very long and instinctively knew, as a responsible parent, that it was time to go back to the cabin when my younger daughter, Elspeth, turned blue. Of course, she was never ready to go. Thermal homeostasis is absent in children.

Almost daily we'd drive some distance down the mountain to return to the wild the mice (Peromyscus) we'd caught in our cabin using live traps. At first we just released them outside our abode at the station, but we were teased heartily by Jerry Wolff and his team of *Peromyscus* researchers, who claimed that those mice would be back inside the cabin before we were. Well, given that that was the case, I figured we should be able to beat them back to the cabin after their release if we were then driving a car up the mountain and the mice were on foot. I'm not sure we did. We didn't really mind the mice eating our soap (which actually happened!), but I didn't want them to chew holes in the bag of my bagpipe. It's crucial for the bag to be airtight, or maintaining steady pressure on the reeds is impossible!

After supper and much animated discussion about various things (as occurred at all meals), there was nearly always a volleyball game, weather permitting. This was the social highlight of the day for my daughters. I took the opportunity to slip away to the other end of the grounds, on the bank of the dam, to practice playing my bagpipe in the relative solitude this special time of day afforded at that location. Pipes are loud, and I didn't want to disturb people. But I had an audience of one nearly every evening—the director of the station, Jim Murray.

I had been a piper for about 15 years by that time, having started at the end of graduate school, when my wife Cathy bought me a bagpipe as a gift to celebrate earning my PhD. I was so excited to get the bagpipe that when people phoned to congratulate me on my thesis defense, I said, "Forget that! I just got a bagpipe!" Cathy is the real musician in the family and has played organ professionally since before we were married in 1964. She still does, and we're still married, despite the fact that I play the bagpipe. I'm strictly an amateur and attempt to play any musical instrument I can get my hands on. I play (or play at) lots of instruments, just for fun. I am no virtuoso on any of them, and it's a long list of short accomplishments. But over the years I became a fairly competent piper. I founded a pipe band in Williamsburg, was a member of two competition pipe bands, and played professionally at Busch Gardens. I mention this here not to sell records, but because it has some relevance later on in the saga. (If you see any recordings by me, don't buy them. I give them away for free. So far I still have all the copies.) In any case, Jim would lie on his back on the grassy slope of the bank of the dam and listen to me practice. This was flattering, but it stifled my attempts to master new tunes. Pipers learn pieces on the practice chanter (a quieter version of the tune-making component, without the drones and air bag) and then, once memorized, they attempt to play the new music on the full bagpipe. It's not always a smooth transition, so one doesn't want to subject an audience to this initial stage. That was the sort of practice I was hoping to get by being well away from the crowd engrossed in volleyball.

During one of my solitary practice sessions, Jim came along, but this time someone was with him. It was his wife Bess. They first met when Jim went to study in England. As Jim tells the story, when I asked him for details in a recent email: "Yes, we met at Oxford, sharing a lab bench as undergraduates, '51 to '54. Then I did my army service, taught for two years at W&L [Washington and Lee], and then went back to Oxford for my doctorate, '58 to '62. We were married while I was in Lexington [Virginia], '57. It took us a while to get used to the idea."

I worry as I write this that some of my observations might be taken as put-downs at the expense of people of whom I am very fond. While there may be individuals I mention in this book for whom I have feelings of reserve, Jim and Bess Murray are not among them. I have considered them as dear friends for many years, and my remarks are offered with affection. I vow not to repeat this disclaimer, however, each time I exploit my friends in humorous anecdotes. My intention is to entertain, not ridicule.

Jim is a tall, gaunt man. He is one of the very best naturalists I know. You can turn over any rock, and he can tell you what's under it, and not just by putting a name to this or that critter, but he can also tell you, in detail and with references (if you want them), what it does and how it makes its living. It doesn't matter if it crawls, walks, swims, or flies, Jim Murray is the man you turn to when you want to know what's going on, anywhere on this planet. He's a quiet man. Thoughtful. Very thoughtful. You almost have to poke him with a stick to force him to express his opinion. Yet he has them, and they're well thought out. He's the sort of guy I would like to be, a sort of Clint Eastwood character in natural history: soft-spoken, reserved, confident. I'd settle for just being taller.

Bess, as Jim described her, was "quintessentially English." Perhaps quintessentially Oxford would have been more precise. Bess, too, was a wonderfully knowledgeable naturalist and a biologist in her own right. But unlike quiet, soft-spoken Jim, Bess was outspoken and didn't beat around the bush for a nanosecond. You never had to wonder what she was thinking because she would tell you, straight away, and she insisted on hearing your views, as well. Between the two of them, they had an encyclopedic knowledge of life on Earth and the best intentions for its well-being as any "save the planet" organization could hope for.

I put down my pipes as Jim and Bess approached, with broad smiles on their faces. Jim, characteristically, was quiet when Bess said, "I want you to play 'Flowers of the Forest' at Jim's funeral."

Jim raised his eyebrows quizzically and then calmly asked her, "Have you a particular date in mind?"

"Well, I hope it's not any time soon," I said, "because I never bothered to learn that tune."

To which Jim replied, "If I hear you practicing it, I'll start to worry."

There were other musical diversions at the station. A number of people brought along guitars, and some could even play them. My daughter Megan played and sang, and she enjoyed those special occasions when people would gather round the campfire or a fireplace in a cabin and sing the night away. This is not to suggest that people didn't work hard, but they also found time to take pleasure in one another's company. While there were organized cook-outs and bonfires, informal, spontaneous gatherings were not uncommon, and most of these happened wherever Charlie Werth happened to be at the end of a long day.

Charlie, who died from cancer in 2001, just as his career was maturing, was a new postdoc when I met him at the station in 1983. To describe him as good-natured or affable would be accurate, but it wouldn't be enough. I was on the porch of the dining hall, standing in the shade and chatting with Jim Murray, when I first laid eyes on Charlie as he walked down the path to the building. He stopped in front of us but didn't come up on the porch. Jim introduced us, generously referring to me as a *Biston* man. Charlie, I learned, worked on the phylogenetics of ferns, using allozymes (enzymes) to work out relationships. We chatted amiably for several minutes, with Charlie still standing in the sun, and Jim and I remaining in the shade. Charlie was wearing a black T-shirt and had begun to perspire quite noticeably. Flies were soon swarming around him, and he nonchalantly would brush them away from the front of his face, talking the whole time. Suddenly Jim asked, "Are you going to just stand out there attracting flies, or are you going to come up here on the porch?"

"Flies don't bother real Mountain Lakers," Charlie said, as he continued to explain his latest results from the lab, his black beard and wild curly hair glistening from sweat. He looked like a bear, a smiling, blissful bear.

Charlie was a field botanist who learned what he needed to know to answer his questions. His interest in working out evolutionary relationships among ferns required mastering rudimentary population genetics and employing molecular techniques. The use of DNA for such work was well underway elsewhere, but Charlie found it more practical for his purposes to use the tried-and-true methods of gel electrophoresis (using an apparatus that applies direct electrical current through a gel to separate molecules having different net charges) to analyze variations in populations at the protein level.

Charlie developed the procedures that worked for him, but his gels were never pretty. At talks he'd show a photograph of what he called a "typical" gel to illustrate a point he was making, but before anyone would ask, he'd admit, "Well, OK, it's not a *typical* gel, it's my *best* gel." I strongly suspect most people show their best gels and call them typical, but only Charlie was compelled to confess.

Yet he worked tirelessly at getting it right, and he shared whatever he learned with anyone who wanted to know about it. In the years to come, he would offer lab courses at Mountain Lake to teach gel electrophoresis to field biologists. He was a lovable character, immensely popular with everyone at the station, both professionally and socially.

I recall a spontaneous cabin party when Charlie got out his guitar and sang the blues, inviting people to join in and make up verses. Most of these centered on their work, and the frustrations with it that gave them the blues. It was hilarious. The next morning at breakfast I overheard a visitor to the station complaining, "I went to a party at a cabin here last night, but when people started singing about their gels, I knew it was time to leave."

There was no shortage of planned entertainment. On weekends we'd have movies, or invited live performances by local folk musicians, and on one occasion there was a station-wide talent show, during which I piped for my younger daughter to dance the fling. (She later became a professional ballerina, and then an engineer. My older daughter still dances, teaches dancing, and is a professional judge at international highland dance competitions. But I digress.) My favorite scheduled activities at the station were the three seminars held every week. Tuesday nights were for formal research seminars, usually presented by invited speakers. On a different night, there would be a less formal talk, by someone in residence, about work in progress at the station, usually with an appeal for suggestions. Sundays were travelogue nights, featuring natural history slide shows of interesting places someone had recently visited.

The summer was over too soon. I'd collected my set of data, such as it was. I had worked hard and made friends with some really nice people. As it turned out, I went back to the station a few more times over the years to teach evolutionary genetics. (I'll discuss that later.) But this first summer ended as it started, on the porch of the dining hall. I was in the shade again, chatting with Jim Murray. I confessed that my data gave me an answer, but I wasn't sure I had it nailed down. I noted that melanism is rare here. Are the specimens I found from a sufficiently polymorphic population of moths? And the typicals are not really pale at all, but dark, so there is much less difference between the pale forms and the melanic forms here than there is between their counterparts in England.

Jim said, "You know what this means, don't you? You'll have to go to England."

I knew that even before he said it, but I was ambivalent about going there to continue this work. It wasn't that England's such a bad place. Not at all. My wife, our daughter Megan (who was only five at the time), and I spent the better part of a summer there in 1972. We rented a small camper van and toured the length and breadth of the United Kingdom. We also visited cousins in Cumbria. We loved it there. My uncertainty was about the practicalities of finding a suitable location to trap moths; having moth traps available; locating a place to set up test pens, a lab with a dissecting microscope, and a source of CO2; obtaining a car to use for several months; renting a place to live; and coming up with the money to pay for it all. I was spoiled by Mountain Lake. It had all been so convenient and easy. Everything was there for me, except the moths I wanted.

I looked out from the shade of the dining hall's porch to see Charlie Werth coming down the path. "What's happening?" he asked.

"I have to go to England next summer," I said, sounding as though it was some form of punishment.

"Well, cheer up, Bruce. You'll like their beer," said Charlie, patting his belly as if he'd just had a pint. We waited expectantly for the burp.

10 | Travel Arrangements

I returned to Williamsburg in late August for the start of classes. Autumn is a very busy time of year, especially for those of us who run labs as part of our courses, and particularly so if living organisms are employed. I had to expand the *Drosophila* stocks, plant corn and soybeans in the greenhouse, transfer strains of fungi, and prepare cultures for my genetics class. The biology faculty at William & Mary prepared and taught their own labs for all upper level courses, unlike at larger universities, where such chores are routinely relegated to teaching assistants. Plus, there are the usual duties familiar to all faculty, whatever their discipline: advising students, getting lecture materials ready, attending meetings, and participating in committee work. Nevertheless, I always looked forward to the start of classes with great anticipation and excitement throughout my career. I had the greatest job on the planet.

Despite being so busy, I immediately whipped up a short manuscript on my Mountain Lake findings and submitted it to *Nature*. Fortunately, it was rejected. The reviewers' comments were trivial, yet sufficient to torpedo the paper. I was disappointed initially, but not surprised, and before the next year was out I was downright relieved. It was not because of any errors the reviewers uncovered, but from what I subsequently learned as I continued to work with peppered moths. My conclusions were correct, yet some of my earlier assumptions, based on previous reports, were flatly wrong, and I'd have been embarrassed to have them in print. All in good time. My brief turn in *Nature* would come.

During autumn 1983, I wrote letters to the three most prominent *Biston* researchers working in England. Jim Murray urged me to write to Laurence Cook, a population geneticist of renown who analyzed field data with mathematical rigor, at the University of Manchester. Jim knew Laurence since their student days at the University of Oxford, and both were products of its storied school of ecological genetics. Laurence worked with a variety of organisms, such as Jim's polymorphic land snails (*Cepaea nemoralis*) and Kettlewell's peppered moths, among other favorites of British naturalists. In contrast, Americans back then mostly studied flies.

Jim also recommended that I contact David Lees at Cardiff University in Wales. Lees, the author of the then most recent review on industrial melanism, had worked extensively with peppered moths. He focused much of his attention on the intermediate forms collectively called *insularia*, a phenotypic category swept under the carpet in general accounts of industrial melanism, although it is a very common phenotype over wide ranges across the geographical distribution of peppered moths.

The third person I wrote to in England was Cyril A. Clarke, who had retired from the University of Liverpool in 1972 but continued working from his home in West Kirby on the Wirral, a peninsula between Liverpool and North Wales. Dave West suggested his name because they had been collaborating on producing hybrids and backcrosses of the American and British subspecies, in attempts to repeat Kettlewell's experiments on the evolution of dominance of the allele (a variant form of a gene, which occurs at the same position on a chromosome) producing the *carbonaria* phenotype. (I'll discuss those experiments and attempts to repeat them in a later chapter.) For many years, Cyril Clarke had coauthored several series of widely cited papers with Philip Sheppard, who died from leukemia in 1976 at the age of 55. Their names were well known to me. What I didn't know yet was that Cyril Clarke wore many hats and had been knighted for his work in medicine. By the time I had written to him he was Sir Cyril A. Clarke, Fellow of the Royal Society (FRS) and Knight Commander of the Most Excellent Order of the British Empire (KBE).

All three gentlemen responded promptly and cordially to my letters, and all of them invited me to come to England to continue the work on my project at their universities. I was immensely pleased by this and now had to decide where to go. It was an easy decision.

In my letters of introduction, I expressed a need to collect hundreds of specimens of peppered moths to test the contrast/conflict idea and emphasized that the project required good representation of both melanics (*carbonaria*) and typicals from the same population. Both Cook and Lees cautioned that it might be difficult to get that many specimens in a single season at their locations, noting that it was more likely the numbers would be in the tens. In the tens?! Well, that was no good.

Sir Cyril's letter was far more encouraging. He had trapped 689 in his garden during the previous season, of which 445 (65%) were *carbonaria*. He said the numbers fluctuated widely between years, depending upon his use of an "assembling trap," which he employed in addition to a mercury vapor lamp trap whenever he had a source of virgin females that could attract (assemble) males by pheromones (a scent triggering a behavioral response in members of the same species). "Do you have pupae [those in the inactive phase between the caterpillar and adult stages of the moth life cycle]?" he asked. While there were no guarantees, he was confident I could get the numbers I needed.

His charming letter said he and his wife lived "in a bungalow, half of which is for people and the other half is for bugs," adding that he had several assistants to help with the work. "Come any time you like but the moths do not start flying until about the last week in May, and their season is over about the first week in August. Our M.V. [mercury vapor lamp] trap has always been in our garden so one can do the catch in one's pyjamas!" Later I learned that he expected me to live in his house.

He added further enticement: "Do you sail boats? I have an 18 footer which I race at weekends in the Dee Estuary and we often need a crew."

Gee, now which place should I choose? Of course I went to Liverpool. All of my anxieties about arranging to work on my project in England evaporated upon receipt of that first of many letters from Cyril Clarke, and it began a collaboration between us that lasted until his death in 2000 at the age of 93.

PART II

11 | Wirral Welcome

I arrived at Liverpool's Lime Street Station on May 15, 1984, and was met by a man bearing a hand-printed sign that read "Professor Grant." He was sent by the Cross taxi service in West Kirby to pick me up. I could have transferred from British Rail to the Wirral Line without leaving the building, but Lady Clarke—Sir Cyril's wife Frieda, or Féo as she was called—worried that I might get lost on my first attempt to reach them. The trip was already paid for, so I hopped in the cab and tipped the driver when he delivered me to the Clarkes' driveway at 43 Caldy Road.

The Clarkes were not at home to meet me, since they had gone to London to attend a soirée. I was greeted instead by a woman named Jean Butler, who went by the nickname Riki. She introduced herself as "Sir Cyril's assistant." She apologized on their behalf for their absence, then showed me around the premises, including the kitchen, the contents of the refrigerator, and my room at the end of a long hallway, which was past the "sail room" (used for mending sails). On the bedspread was a handwritten note for me from Lady Clarke. It read "Welcome!"

It was well past suppertime, so I apologized to Riki for causing her to have such a long day at work. She assured me that this was a quite normal part of her duties, as she had to come there at all sorts of hours to open or close greenhouses or turn on moth traps or do whatever was necessary when the "boss" was away, so there was no need for me to apologize. She explained that the Clarkes were in constant demand and were always on the go. They kept an apartment in London to stay close to the action, returning home to West Kirby on weekends.

Riki was all business: very formal, proper, efficient, and perhaps just a little nervous. She took her work seriously and was clearly devoted to the Clarkes. I thought she was about my age (42 then), maybe a few years older. I hate guessing women's ages, because of possible offense, and I am so bad at it. I probably didn't even think of asking at the time, and if I had, I'm sure Riki wouldn't have given it a second thought. She was totally unpretentious.

As she was about to leave for home, I asked her if we might set up the MV (mercury vapor) trap to see what was flying. It was quite cool that early in the season at that latitude, but I didn't want to miss any moths. Cyril normally ran his trap every day from the first day of June until the end of July. I couldn't wait that long. I started to dismiss the idea, realizing that I was keeping Riki from her family, but she came alive with excitement at my request. It seemed she was hoping I'd want to use the trap right away. She had just installed a new plastic funnel that day, in anticipation of my arrival. Long exposure to the sun over the years had destroyed the old one they had been using. It's a bigger job than it sounds, because she had to fashion the new funnel from a flat, wide plastic sheet and attach it to a metal frame by sewing it on by hand. It had to fit snugly, so any moths caught inside couldn't escape. The two of us rolled the big metal cylinder onto the lawn, placed some egg cartons along the circular wall inside, capped it with Riki's newly refurbished funnel, and switched on the mercury vapor lamp. Riki was smiling from ear to ear. So was I. We were friends from then on.

After setting up the trap, Riki went home. I stood in the Clarkes' garden for a while, basking in the brilliance of the mercury vapor lamp, admiring the manicured garden and the ambience of their home, and recalling how I came to be there. I considered myself very fortunate to be about to collaborate with a living legend in science. Then I went for a walk and found a pub.

At the pub I knew what to do. I had been to England previously and learned not to ask a bartender for a glass of beer. He wouldn't know what to serve me. Precision is required. One must specify the quantity and the type. I don't necessarily mean a brand name, although that would be the simplest thing to do. As I was unfamiliar with the local brands, I'd have to say something like, "A pint of mild, please," or "Half o' lager." There's a long list of possibilities, but I was well practiced in ordering "A pint of bitter, please," taking care not to pronounce the double *t* as a double *d*. Asking for a pint of "bidder" would instantly identify the speaker as an American. The *r* in that same word presents another subtlety, as there are regional differences in its hardness. Some Brits sound like they're saying "please" (*bitte*) in German when they order bitter.

I wasn't ashamed to be an American, especially back then when everybody loved us, most especially in England, or so I thought. But I wanted to avoid the inevitable conversations that typically began with, "Oh, so you're an American, are you? How do you like it here?" On and on this could go. Mostly people were just trying to be polite and make visitors feel welcome. Some were genuinely curious and meant no offense. None taken. I just wanted to spare them and me that conversation. I preferred to blend in and be ignored. My behavior was more mimetic than cryptic.

After the bartender greeted me with "Yes?" I said, "A pint of *bittah*, please." He drew the pint, set it down in front of me, took my money, and said, "Ta." Good, I thought, he didn't suspect a thing. Just as I reached for the glass one of the two men standing near me at the bar asked, "Where are you from?" Clearly my phony English accent needed finer tuning.

We swapped a few tales and a few rounds. Over time I told them I was a bagpiper, in case they should happen to know of any in the area. My hope was to find a suitable place to practice or to hook up with a pipe band. I planned to stay in England until the end of August, so I brought my pipes along for company. There was some head bobbing, including the bartender's, as they agreed there was a local piper who stopped by occasionally, but none of them could recall his name. I left the Clarkes' phone number, should the chap come by again.

I was still wide awake when I let myself into 43 Caldy Road, so I went out back to check on the trap. Its glow was clearly visible, even from the road, but I wanted to see if any moths were swirling about. Not much action, so I went inside and thought about bed. It had been a long day, and I should've been exhausted. As I walked down the hall leading to the guestroom, I noticed for the first time a lot of framed photographs on the walls. I stopped to look at them and ended up doing so for several hours. I also glanced through a few books and read some of the inscriptions, including one from Kettlewell himself to a fellow physician. They were not just colleagues; they were close friends. Cyril Clarke wrote Kettlewell's obituary, concluding with "Everyone loved him."

One book in particular caught my eye and held my attention—*Genetics for the Clinician*, a pioneering work in medicine written by none other than Cyril Clarke in 1962. Today, medical genetics is a major industry, but in 1962, an understanding of the role of genes in human health was in its infancy. Cyril Clarke is best known for his role in the research leading to the prevention of

Rh (Rhesus) incompatibility (named after the monkeys in which the antigen was first discovered). This hemolytic disease (causing the destruction of red blood cells) occurs in newborn infants with an Rh-positive blood type whose mothers are Rh-negative. Fetal cells can leak into the maternal blood stream across the placenta. It is imperative to destroy those incompatible cells before they trigger the mother's immune system into producing massive quantities of antibodies that can result in the death of the fetus. To prevent this, pregnant Rh-negative women are injected with antibodies that destroy fetal Rh-positive cells in her blood stream. If the maternal immune system has already been sensitized, however, the preventive therapy doesn't work, so other measures to save the baby (e.g., blood transfusions) must be pursued. Cyril Clarke was knighted in 1974, primarily for this work and for studies on other inherited afflictions in humans.

It soon became obvious that I had been exposed to just one side—my small side—of Cyril Clarke's manifold interests. The pictures on the wall didn't tell the whole story, and I would soon learn more. In addition to a number of photos showing Cyril Clarke being presented with honorary degrees or medals in recognition of his accomplishments, he was also in posed group shots with other recipients of various awards. I wasn't always sure what the occasions represented. Perhaps his knighthood, or installation as a Fellow of the Royal Society, or his inauguration as president of the Royal College of Physicians? But he always looked quite at home among his peers, in each photograph at every event. There were also images of him with his wife Féo. They were a strikingly handsome couple. And there were action shots of them sailing. It was his pastime and passion. He raced sailboats competitively and nearly always won. Who were these extraordinary people? I wondered as I finally fell asleep.

I was up bright and early the next morning to empty the trap and take a shower before Cyril's other assistants arrived for work. I felt amazingly good, considering how little sleep I'd had and that England is five time zones ahead of Virginia. I had made myself a cup of instant coffee and was reading a newspaper when I heard someone enter the house. It was Winifred Cross, Cyril's senior assistant.

She had worked for him the longest and was approaching retirement age. I would not have guessed it from her youthful appearance, as she was very vibrant, energetic, talkative, and keenly interested in everything. Again, I probably did not think about her age when we met, but later she would tell me she had dated an American serviceman who was stationed in England during the war (meaning World War II). She even asked me if I'd look him up when I returned to the States, just to extend her regards and see how the years had treated him. (I was just a baby during World War II, so I would have no "before" to compare with the "after.")

Nothing of this sort was discussed during our initial meeting, but over the course of that first summer—and the several others when I returned to England—we talked at great length about many things of mutual interest. I enjoyed her company and, later, the company of her husband Mick. On my first morning at 43 Caldy Road, Win, as she was called by everyone except Lady Clarke (who always referred to her as Winifred), entered the room with a broad smile and a cheery greeting. "Hello! You must be Professor Grant! I see they've got you working already," in reference to the moth trap set up in the garden. "Any luck?"

"No, but it was pretty chilly last night, so I didn't have high hopes."

"Yes," she said. "And it is a bit early. We usually don't begin until June."

Initially I was unaware that Cyril, at one time, had begun trapping weeks earlier in the season. His daily collection records date back to 1959, and I later analyzed them, along with the whole series of years through 1998, to investigate intraseasonal variation in emergence times of phenotypes. I also graphed his 1960 daily catch records to illustrate the univoltine (single generation per year) pattern of emergence, beginning in mid-May, to compare it with the bivoltine (two generations per season) flying pattern of peppered moths in Michigan during that same year. In any case, the peak emergence in West Kirby comes in mid-June, so Cyril didn't miss many when he decided to not bother trapping in May or August. (We'll return to this point later in our story.)

After chatting a bit more, Win showed me what they were up to in the greenhouses. There were two in the back garden. Cyril was always on the lookout for wing-pattern variations in butterflies from all over the globe, and he had to import appropriate food plants to raise them. He was still actively pursuing research he'd begun with Philip Sheppard decades earlier on the genetics of wing-color polymorphism in swallowtail butterflies, and just then he was investigating distorted sex ratios in *Papilio nandina*, a naturally occurring hybrid from Kenya. In addition, he maintained a project on sex determination and the controversial issue of fertility versus sterility of interregional hybrids of gypsy moths (*Lymantria dispar*). A few summers later, he'd rediscover Sheppard's "lost colony" of scarlet tiger moths (*Panaxia* [= *Callimorpha*] *dominula*), famous from earlier studies comparing the roles of genetic drift and natural selection on wing-spot patterns. Whatever his latest interest, he pursued it with exuberance, and livestock related to the project were labeled VIP to remind his assistants to take special care. VIP labels were all over the place, some old, some new, marking Cyril's long trail of enthusiasms. Win would later tell me that Sir Cyril and Lady Clarke were "almost childlike in their wonder," something most of us lose but continue to envy.

Win introduced me to a freshly minted PhD who was finishing projects of her own and helping out in the greenhouses. I made some comment about what a busy place this was with so many projects under way, to which she replied, "We'll put you to work." I'd hoped she was just joking, because I had my own plans.

Soon I'd meet Angela Urion, the youngest of Cyril's permanent assistants. She had come to work for him as a teenager, I think the story goes, and had been with him her whole adult life. She did most of the clerical work—typing manuscripts and correspondence—but she was fully versed in the other duties all of his assistants shared, including spreading wings to display the patterns of pinned specimens of butterflies and moths.

All of Cyril's assistants were competent at emptying the trap and identifying phenotypes. They also took turns covering for each other when called upon, although, under normal circumstances, they each had primary responsibilities. Riki, for instance, handled the peppered moth duties, and that particular summer she was busy organizing a survey in North Wales by recruiting farmers to run moth traps on their property. Sounds easy, doesn't it? Win's forte was working in the laboratory and photographing specimens. She had even authored a paper on a technique she'd developed to identify the Smith bodies (sex chromosome material) of gypsy moths. Angela was the secretary. Whenever needed, however, it was all hands to the pump. They had been together for so many years that they worked like the infield of the New York Yankees. And they enjoyed each other's company, never gossiping about one another, at least never to me. They were formally employees of the University of Liverpool, but they were paid through Cyril's grants from the Nuffield Foundation. I was surprised that Cyril had such an active research program, as he had been officially retired for a dozen years and was then 76 years old. I was also impressed with his team. I wondered if he knew how lucky he was to have such a dedicated, highly competent staff. Maybe he was the reason for it.

I took the rest of the morning off to find myself a place to live. Although the Clarkes very generously invited me to reside in their house as their guest for the entire summer, I did not want to intrude on their lives. Their offer was not entirely altruistic, as they hoped my presence would keep burglars away while they were in London during the week. Their house had been a target more than once, after which they installed an alarm system that had to be switched off with a key immediately upon entering. It seems as though every home in England gets burgled as a matter of course. At any rate, I preferred to have my own place to live for three months. I had in mind Benjamin Franklin's comparison of houseguests and fish stinking after three days. Besides, once public school in Virginia let out, my family was to join me for the rest of the summer in Merry Old England. We'd be camels in the tent at 43 Caldy Road.

After perusing the classifieds and making a number of phone calls, I found a furnished flat that could be rented for just three months. Most apartments required longer-term leases, so options were few. I made an appointment to see the place the next morning. Unfortunately, it was located in Birkenhead, across the width of the Wirral, several miles away. Taking the train to work every day, especially carrying cages of live moths, was impractical, so I decided to see about getting a car for the summer.

After a few more phone calls I concluded that the expense of hiring a car for three months was a lot more than buying a used car, depending on how old and how used. I walked to a nearby car dealer in West Kirby and talked to a salesman about what he might have in stock. He asked my price range and laughed at my answer. "What you want is a banger," he said, and he advised me to check the local papers.

There it was: "Hillman Hunter, good runner, £100." The British pound sterling then was equivalent to roughly $2, give or take, so I immediately rang the number to set up a test drive. I took the train to Hoylake and walked from the station to the address. The banger was parked outside, painted yellow and black. Warning coloration, I thought. The young man who was the seller was a fast talker and an even faster driver. He ripped up and down the road in the Hunter to show that it was indeed a good runner. He asked me if I wanted to drive it, but I declined, thinking that later and slower would be better. He also explained, much to my satisfaction, that the M.O.T. (Ministry of Transportation) sticker on the windscreen wouldn't expire until after I went back to America, so there was no need for more fees or vehicle inspections. Good deal, I thought, and said, "So, it's £100, is it?" He said, "No. You must have seen the earlier advert. I dropped the price to 85 quid." What an honest bloke, I thought.

We exchanged my money for his keys and shook hands. As it was past 5 PM, I would have to take care of the legal paperwork the next morning, including getting insurance. Should I leave the car parked there, or take a chance driving it back to West Kirby without insurance? The seller encouraged me to drive it away. (I assume he just wanted it out of his sight, or maybe his neighbors had put pressure on him to remove it from theirs.) "It's a straight run, not far. Just keep to the left." I knew that. I had driven all over the United Kingdom in 1972 without mishap. As long as I pay attention to traffic and concentrate on keeping left, I thought, it'll save time and running around tomorrow. Then I opened the door on the passenger's side of the car. I realized my mistake immediately when I didn't see the steering wheel. "Oh, right," I laughed, "other side." No time for second thoughts now. I was committed, or should have been.

The fuel gauge read empty, so the first thing I did was to stop for gas (err, petrol). The petrol station was on the right side of the road from the direction I was heading. What a good deal this is, a car that actually runs well for only £85, I kept thinking. How lucky can a guy get? I paid for the petrol, hopped back in the car after first opening the door on the passenger's side to once again be certain no steering wheel was there, then looked down the road to make sure nobody was in the left lane, the lane I was about to enter. Nothing was coming for as far as I could see up the road, so I stepped on the accelerator and let out the clutch.

Bang! I heard the impact and saw the blur of a car shoot past simultaneously. Just as the nose of the Hillman Hunter, my banger, went out into the road, a car coming from the opposite direction, being driven properly on its left (but on my right), made contact. It wasn't much, but it was real, and it was entirely my fault.

The other car screeched to a halt and the driver flew out, in a rage. I couldn't blame him. I was furious with myself. It wasn't the damage that upset him. It was very minor to both cars. He kept repeating that I pulled out right in front of him and if he'd had his children with him they could've been killed. He was right, and I could do nothing except acknowledge the fault and repeat my apology. We then carefully examined both cars for damage. His had a minor scratch that, unfortunately, ran the entire length of his car. The Hunter didn't look noticeably much worse than when I bought it, except the left wing (fender) was openly separated from the bonnet (hood) and drooped away from the body. When we got to the part about exchanging insurance information, the other driver once again exploded when I told him I didn't have any. "What?! I'm calling the police! You Americans think you can come over here and do anything you want and get away with it."

I assured him that I had better intentions, was going to get insurance first thing tomorrow, had just bought the car and was hoping to sneak it home, and so on. He calmed down after I promised on my word of honor that I would pay for repairs to his car. I gave him the numbers from my passport, my driver's license, and Cyril Clarke's phone. We shook hands after he reluctantly agreed not to call the police. Then we drove off in opposite directions. When I got back to 43 Caldy Road without further mishap, fortunately everyone had gone home. I parked in the driveway, found some old wire in the garage, and fastened the Hunter's wing back on. I had learned a valuable lesson on my first full day back in England: jetlag is real. I was lucky the accident didn't cost me or someone else a whole lot more. Then I switched on the moth trap and went to bed.

The next morning I rang an auto insurance agent to set up a policy. Then, after the necessary running around about car registration, I went to Birkenhead to check out the apartment. It was in a large, brick, Victorian house on the main drag. I parked my banger on the left side of the road, in the direction of Liverpool, and walked across four traffic lanes to the house. The door was locked, so I waited a few minutes for the owner to arrive. He drove up in an estate (station) wagon, coming from the same direction I had, but he crossed the road in his car—all four lanes—and parked facing in the wrong direction to the flow of traffic. Watching this made my heart race, but he did it very casually and was quite relaxed when he exited his vehicle to greet me. His wife normally handled the rental business, and he did the repairs, but he happily showed me the premises. There were apartments on all three floors, and mine was on the first floor. It was the only one with a private, fenced-in garden. It was wildly overgrown from neglect and very small, but it was the deal clincher as a perfect place to set up moth pens. "I'll take it," I said. The landlord asked if I had pets. "No," I said. I didn't consider moths to be pets.

Before the day was out, I had moved in. For the rest of the week I'd drive to 43 Caldy Road to turn on the trap at dusk and then drive back before 6 AM the next morning to switch it off and examine the night's catch. Getting there at dawn wouldn't allow me much sleep at all, because summer nights are very short at that latitude. It was still May, but the days were getting longer. The ride itself was very pleasant along the route I had picked out, and I slowly regained confidence in driving on the left. Actually, that part is simple once you are in the proper lane. It takes a little more concentration when making turns at intersections or negotiating roundabouts, but soon I was driving like a native. Now I drive on the wrong side of the road only when I forget which country I'm in.

My first meeting with the Clarkes occurred on a bright, sunny morning when I arrived to switch off the trap. I knew they would be back from London for the weekend, and I wanted to be sure to get to the trap before Cyril did. After pulling out the plug and waiting for the bulb to cool, I removed the funnel and began to unpack the trap. "Hello," I heard Cyril say from an open window.

12 | Coffee with the Clarkes

Cyril's greeting came from his bathroom window, as occurred uncounted times over the years I knew him. He would push out the casement, boom out a "Good morning" or "Hello," and then get right to business. "Any *betularia*? How many *carbonaria*? What do you have?" These are not exact quotes from each morning, but, basically, he wanted to know the score. And he wanted to know it the very first thing each day. Sir Cyril Clarke did not beat around the bush, or, if he did, his beating was cursory at best. He didn't waste time on small talk. That first morning was a little different, as we hadn't yet met. "I'll be right there," he said. I could hear the water running in his lavatory as he spoke.

I do not mean to suggest that he was rude. He was never rude—at least not to me. He could be brusque at times and express his impatience with people, but typically he was the consummate gentleman, practically typecast from a Rex Harrison film. He could charm the pants off a brass monkey. (I'll return to his charm later.) My first meeting with Sir Cyril was quite casual. He came into the back garden in his dressing gown (bathrobe), shook my hand, and handed me the latest issue of the *Biological Journal of the Linnean Society*, turned to a paper by the Finnish lepidopterist Kauri Mikkola. "Have you seen this?" Cyril asked. "No," I answered, embarrassed to not know what was going on. How could I have missed that article? It must have come out while I was on the airplane! Cyril excused himself to get dressed, leaving me to read what Mikkola had written.

Woe is me, I worried as I read. I've been scooped! No, as it turned out. Mikkola's interests were very much in the same ballpark as my own, but he wasn't asking the same questions, and he certainly did not test Kettlewell's contrast/conflict hypothesis. I was relieved to find I still had a clear field all to myself, having read the very latest stuff on the subject; heartened to know that people still cared about the problem; and pleased that it was published in a major journal.

While I was still reading Mikkola's paper, Lady Clarke appeared in the garden, bearing a tray with both coffee and tea and proffering an enthusiastic invitation to join them for breakfast. I stood up, bowed, and smiled, but declared that I'd already eaten. I probably fibbed about that, as I did not want to intrude on their lives, not even for a moment. I did agree, however, to join them for coffee after they'd had their breakfast and suited up for the day.

Just as I was finishing Mikkola's paper, very relieved that I hadn't been scooped, Cyril returned to the garden to usher me inside for coffee and conversation. His dressing gown had been replaced by a baby-blue pullover sweater and beige trousers. His pure white hair was combed back and parted smartly above his right eye. He looked like a knight, not in shining armor, but one of notability. He asked, in direct reference to Mikkola's paper, "What do you make of it? Anti-Kettlewellian, is it?" Then we went inside for coffee and began a dialogue that lasted for 16 years.

I didn't know anyone other than my own mother who could've made me feel so at home and welcome as Lady Clarke. She asked me to call her Féo, but in deference to the difference in our ages, I couldn't do it. The same went for Sir Cyril. For me to have called him just plain Cyril, sans the "sir," was to presume, on my part, that we were peers. Clearly we were not. This is not to suggest that in any way did Sir Cyril and Lady Clarke fail to treat me fully as a peer and a colleague, a fellow professional. The problem came from my own upbringing. Neither of my parents, children of immigrants from Europe, had graduated from high school. My father, born in England, was brought to America when he was nine years old and put to work in the coal mines of Pennsylvania when he turned 14. His mother, widowed by the Great War (World War I), had a lot of kids and needed money to feed them. School was a luxury for them. Still, my father became well educated through his own efforts. He read voraciously, especially works on history and mythology. He also read and wrote poetry, knew art and enjoyed painting, sculpted wood, and loved classical music and opera. He was also a superb craftsman. "Good enough" was not part of his vocabulary. He retained an Old World formality. Kids and strangers didn't call him Bill! As an adult he was "Mr. Grant," even to his boss. When he lay dying in a veterans' hospice, nurses would call him "Bill," and I would wince. There was no way I could address Sir Cyril and Lady Clarke informally. I was a peasant, and they were nobility.

"Do you take sugar?" Lady Clarke asked politely as she shuffled and clanked the cups and pots. She then chuckled, embarrassed by her faux pas: "First we should establish if you prefer tea or coffee to put sugar." She was animated and sparkling, and the words "prefer" and "put" were stressed playfully by increased volume and rising pitch.

Cyril said, in a deep voice, "Yes," dragging out the word "yes" in exaggerated agreement. This rekindled a memory for him. "Did you know Fisher?" he asked, referring to Sir Ronald Fisher, the mathematical genius who invented so many of the statistical tests still in use, including the analysis of variance. He was also one of the founders of population genetics and author of the fundamental theorem of natural selection.

"Of course I know who Fisher is," I grinned, "but I never met him personally." Sir Cyril knew him well enough to write his obituary, and he had this story to tell about how Fisher came up with his Exact Test (used to determine whether there is a significant association between two variables).

"Should you put sugar into the tea, or pour the tea onto the sugar?" Cyril began.

"Or is it coffee?" Lady Clarke interrupted.

"Doesn't matter whether it's tea or coffee," said Cyril.

"Well, it does to the person drinking it," laughed Lady Clarke.

"It wasn't sugar at all," insisted Cyril, "It was milk or cream or something. The point is, can one tell the difference if the tea is first poured into the cup and then the milk is added to it, or the other way round. Put the milk in first, then pour the tea on top of it?"

"Yes, I know this story," said Féo.

"Do you know this story?" Cyril asked me.

"No. Please continue," I begged.

As the anecdote goes, Fisher was at a social function at the Rothamsted Research Station and offered a cup of tea to a staff member. She declined it, explaining that the milk should have been poured into the cup first. Fisher challenged her, stating that surely it made no difference. He devised a test to see if she could pick out those cups that had the milk put in before the tea was poured, from the cups that had the milk added to the poured tea. Fisher apparently was fond of the story himself in relating how he came up with his Exact Test, but he left out the part that the woman actually could tell the difference between the cups of tea that were prepared differently. Cyril very much enjoyed relating what a rascal Fisher was for not giving the woman credit for her discriminating palate but admitted that Fisher's daughter set the record straight in her biography of her father.

"Yes, yes, she could tell the difference," said Féo. "So, do you take sugar?" she asked me. "We'll sort out what to pour into the cup later."

They were quite at home, and they made me feel the same way. It was easy from day one. All of my apprehensions about meeting and collaborating with a living legend in science and medicine evaporated when Lady Clarke poured me a cup of coffee that first morning. Or was it tea? I no longer remember. I do recall that I took sugar back then. They could've served me Bovril and I'd have been just as happy.

There was the inevitable chitchat, but it didn't last long. Most of the obligatory polite inquiries came from Lady Clarke, such as "Are you settled in?" Some of the discussion focused on their sincere invitation that I stay with them, with my polite (I hope) insistence that I needed my own place, because my kids would join me later that summer. At this point Cyril hopped back into the conversation, stating, "We wondered about that." From my letters, they had speculated that I might be a single parent, separated or divorced, with children to entertain during the summer. I assured them that was not the case. My wife would be coming for some period during the time I needed to be in England, so we'd require a full-sized apartment. Most of the chitchat was controlled by Lady Clarke, who was very warm, very cordial. Sir Cyril smiled affably throughout this, then, seemingly out of the blue, asked, "Do you have a car?"

"Yes," I said, but resisted the impulse to add, "It's dripping oil onto your driveway as we speak." He was happy with the short answer and laid out plans for me to go to North Wales to bring in catches of peppered moths. I initially balked at this suggestion, as I had plans of my own. Knight or no, I wasn't about to become Sir Cyril's squire. I spelled out my conditions if I was to go to North Wales. I said as politely as I could manage that I didn't want to retrieve dead specimens for census-taking purposes, but if I could get freshly caught, live, healthy moths that I might use in rest-site selection tests, I'd be happy to go to Wales to collect them. I could then compare the behavior of moths we caught here, near Liverpool, with that of moths living in the relatively pristine conditions of North Wales, where melanism in peppered moths had been historically low.

I went on to ask if extending his survey of melanic frequencies into North Wales was part of the national survey being organized that year by Laurence Cook, through the Open University program. Cyril scoffed at the idea, saying that his colleagues, for whom he had the greatest respect and affection, "act as though they have all the time in the world, but I don't." He eagerly recruited coworkers and coauthors to contribute to his papers, but he marched to the beat of his own drum.

I did go to Wales a few times that first summer, with a caution from Cyril that the Welsh don't like the English, but that I shouldn't have any problems since I was an American. He smiled in amusement as he said this and added, "But I love the Welsh anyway!" As it turned out, Riki handled the Wales survey, and I went down only when large, live catches were available for my work and she wouldn't be returning to West Kirby that day. Riki and her husband Tony had bought a cottage there and intended to retire in Wales. Tony was an art teacher and painter, and he was a great fan of J. M. W. Turner. I was a blockhead when it came to art, but I enjoyed visiting them at their cottage on several of my brief trips to Wales. I also retrieved moths from Mr. Goronwy Wynne, a science teacher at the North East Wales Institute of Higher Education in Wrexham. My trips through the picturesque countryside were always enjoyable, with one-lane roads in the mountains, old stone bridges, and sheep dotting the landscape. Riki arranged her vacations around lambing time, so she could assist the local farmers. I once asked Lady Clarke about Riki's involvement with sheep, to which she replied, "Yes, Riki loves sheep. Absolutely loves them! But she doesn't actually own any." Her characteristic way of emphasizing words using volume and pitch boomed strongly on the word "own." This was not offered in derision, but in sympathy, as if to say, "Poor Riki."

Before ending our first meeting, Sir Cyril asked me to pay a courtesy call to the head of the genetics department, Donald Ritchie, at the University of Liverpool, where I was officially a guest researcher. I rang his secretary, Denise O'Leary, to make the appointment. I recall her name because Cyril acknowledged her considerable help in providing him with information for Philip Sheppard's obituary. She had been Sheppard's secretary. Continuity. My interview with Professor Ritchie was pro forma. The department, like so many others in the biological sciences, was "going molecular," and its former strength in ecological genetics was a distant memory. Still, Cyril Clarke had brought so much money into the university through the Nuffield Foundation—and continued to do so even in retirement—that he was held in high esteem. They kept his numerous awards and trophies on display.

During my brief chat with Professor Ritchie, I noticed he was a smoker. As I had been in the process of quitting smoking for several years, I was tempted to bum a cigarette from him but thought better of it. By that time among American biologists, smoking was a fading habit, yet it was still surprisingly common in Britain. We grew the stuff, and they smoked it. Allies. Ritchie asked about my needs, and I told him I required CO_2 to anesthetize moths. He took me to their *Drosophila* lab, where they fixed me up straight away: tank, regulator, hoses, trolley for transport. All free! We loaded the stuff into the Hunter and I drove fearlessly into Liverpool city traffic. This time I had insurance.

Back at 43 Caldy Road, I moved into the small lab attached to the Clarkes' one-car garage. Its front room was filled with gear to rear specimens and cages for moths and butterflies. There was also a toilet. Beyond that was the small room Win had used for her lab. They pretty much turned all of it over to me for the summer. It had almost everything I needed, including a dissecting microscope and, now, CO_2. The missing ingredients were some live British peppered moths, black and white. Maybe tomorrow?

13 | *A Clockwork Orange*

Very few moths of any kind came to the trap during that first week, so I had a lot of free time to take long walks, explore the area, read, or play the bagpipe. I'd given up on the idea of practicing in my flat, because the pipe is loud and the walls were thin. It was an old Victorian house with high ceilings, but I could plainly hear the young woman on the top floor screaming in ecstasy for about 20 minutes nearly every afternoon. Occasionally I'd encounter her coming down the stairs, always hurrying and looking embarrassed, followed by her husband, looking belligerent. In any case, I knew that sound traveled throughout the house, so I did my piping elsewhere, out of consideration for my neighbors.

Conveniently for my purposes, the house was a stone's throw from Birkenhead Park. Someone told me it was the prototype for Central Park in New York City. I would not have guessed that from its size and appearance, but the park was big enough to include quiet nooks to "strike in" bagpipes without creating a disturbance. I preferred to practice without an audience, but on my first outing, several teenage boys gathered around to listen. They looked like the cast from *A Clockwork Orange*, so I was a bit wary. As I played, they listened politely, and I became hopeful that perhaps they might provide the link I was seeking. Maybe they were pipers, or knew pipers? After all, this was Britain, and Liverpool was the major seaport through which emigrants left the country, including masses of Scots and Irish, many of whom only made it as far as the seaport cities and settled there for economic reasons (no money for a ticket out). Birkenhead is separated from Liverpool by the Mersey River, so maybe these kids could help me hook up with some local pipers.

I stopped playing to make some adjustments to the tape on my chanter, and one of the kids took the opportunity to ask me a question. "Did you make those?"

"Make this bagpipe?" I asked incredulously, as my hopes that these kids could help me vanished with his question. "No. I didn't make them."

I didn't say very much more before one of the other kids asked, "Are you an American?"

"Yes," I admitted.

"You're an American, and you play bagpipes?" They seemed very surprised by this, so I decided to have some fun.

"Sure I play the bagpipe. Lots of Americans do. Haven't you ever seen any American movies? Cowboy films? Westerns? Haven't you ever seen a lonely cowboy on his his horse, riding off into the sunset while playing a bagpipe?"

"No! Is that true? You're joking. Aren't you?" They didn't buy it, but I stuck to my story.

"I'm not making this up. It wasn't until Gene Autry and Roy Rogers started playing guitars and singing that bagpipes faded out of the picture."

"Well, he is an American, and he is pretty good," a kid conceded.

"Thanks. Ta." I was happy for the compliment. "Do you boys know any pipers around here?"

"No. You're the only bagpiper we ever met."

They were curious about how a bagpipe works, so I pulled the chanter out to show them its oboe-like double reed; removed a drone from its stock to display its single-tongued reed; explained how the bag served as a reservoir of air, allowing four pipes to sound simultaneously; and demonstrated how the pipe is tuned. I concluded the lesson by playing a few tunes for their entertainment and my much-needed practice. Then I went home, thankful they didn't beat me up and take my bagpipe. They were nice kids, once you got past their frightening appearance. The punk look started here, and they presented advanced examples, even by English standards. I pretended they were normal. They probably were. They weren't so sure about me, but they applauded and hooted approvingly when I finished with some peppy jigs, just to show off.

A few days later, after switching on the trap in West Kirby, I decided to go for a walk around my neighborhood in Birkenhead. It was a dreary, gloomy evening, and peppered moths were still not flying. It was too early in the season to be concerned about that, but I was lonely, with too little to do. West Kirby, especially along Caldy Road where the Clarkes lived, is a fairytale section of England, with old stone buildings in the villages, stone walls lining winding roads through woodlands and farms, and sailboats moored in the Dee Estuary. In utter contrast, Birkenhead, on the Merseyside of the Wirral, is grim and grimy. Every street is endlessly the same as every other one, lined with austere houses jammed together with such similarity that it's easy to become lost. The pavements (sidewalks) were covered with dog poop, and the familiar red pay-phone booths all stank from urine. Finding a pay phone that worked was a rarity. Most had been vandalized or looted. In the early evening the streets were relatively quiet, with scattered shoppers and commuters returning home from work, but by nightfall the "changing guard" included loud drunks, purple-haired punks, druggies, and yabbos. Still, I felt much safer than I ever did when walking inner-city streets in America after dark. Here I might get beaten to a pulp, but it was very unlikely that I'd be shot to death for my wallet.

The sound of an alarm penetrated my consciousness. It was coming from a nearby house. Just then a police car came straight toward me and the house with the sounding alarm. Now we'll see some action, I thought. Here the cops actually pay attention to such alarms, I thought. But no, the police car just continued past me and the house, completely ignoring the loud ringing as they went on about their business, whatever that was. Surely they'd heard it. I couldn't help but be amused at how secure their home alarm made the Clarkes feel. Apparently house alarms are just as effective as car alarms—more annoying than alarming. At least that's how the police view them on both sides of the Atlantic. I decided to keep on walking so I wouldn't get arrested, just in case the cops came back. I wasn't walking very fast, because a guy carrying a television set went right by me.

Several blocks onward, I caught a fleeting but familiar image out of the corner of my eye. Bagpipe drones! Way down the street I spotted a man in a kilt climbing into a Volkswagen Beetle. I could see the cords and tassels swinging from the drones of a bagpipe as he pulled them into the parked car. I yelled "Hey!" and kept calling out as I ran full throttle down the length of the street to stop the car before it pulled away. I barely made it in time. The vehicle was just leaving the curb as I ran alongside, shouting into the window. It stopped. There were two guys inside, both wearing kilts. The passenger cradled his bagpipe in his arms, and the driver's pipes were across the back seat. They look at me quizzically. "What?!" They demanded.

I was out of breath from my mad dash to catch them. I blurted out, "I'm a piper!" They seemed to understand immediately why I would chase them down the street. They told me they were late for a gig, so they were in a hurry. They did mention that they would have band practice the next night. They wrote the address on a slip of paper and sped away. I was no longer lonely. I had mates!

The address was in inner-city Liverpool, for a recreation building that belonged to a small church. The band was not affiliated with the church in any way; it simply rented the facility. This is a common practice. Churches need money, and pipe bands need practice space. It's an ungodly alliance. I didn't want to be late for band practice, but I had to travel from West Kirby after switching on the moth trap. The band had already tuned up and was playing a set when I entered the room. It was just large enough to squeeze in a full basketball court, if you didn't mind crashing into the walls when going for layups. The pipe band was circled up in the middle of the floor, and when I walked in everyone took the cue to cut the tune at the end of the measure. They all turned around to great me with broad smiles of welcome, but the only person who spoke was the man clearly in charge. I'd assumed he was the pipe major, but he assigned that title to one of his students and referred to himself as the band tutor. Whatever he called himself, he was, in fact, the pipe major. "Welcome," he said with great enthusiasm. "We were expecting you. I'm Tom Graham."

Tommy was a man of wiry build, short stature, and fiery disposition. He sported a close-cropped, full beard and looked to be in his early 60s. I am no better at guessing men's ages than women's, but I learned that Tommy had been, for a time, the youngest acting pipe major in the British Army during World War II, serving with the Argylls, a Highland regiment. He was clearly a lot older than the other band members and had grown children, two of whom still played in his band. In any case, I was roughly 20 years his junior. I told him my name and said that I was a visitor in the area for the summer. He then introduced me by name to all of the other players.

The two pipers I'd met in Birkenhead the previous night, Joe and John, were both PhD students in chemistry at Liverpool Polytechnic (now Liverpool John Moores University). Sean, a tall, handsome, clean-cut young man, was the titular pipe major, and his attractive girlfriend Pip was a side (snare) drummer. Another side drummer was a fellow named Harry, and a burly bloke called Big Kenny rounded out the drum corps on bass. Les was an older gent, not quite Tommy's age, who talked very softly, with an incomprehensible Scouse dialect. Les played the pipes, but his first love was American country music. Then there were Tommy's two sons, Andrew and Gordon. Gordon doubled on drums and pipes, whenever needed. He was good at both. He was also university bound, and Tommy was justifiably proud of him. Gordon had already acquired the Oxbridge lisp (an affectation characteristic of socially supraliminal people who've gone to Oxford or Cambridge) in which the letter *r* is pronounced like *w*, so words like "road" came out as "woad," or my name, Bruce, is "pwonounced" as "Bwoose." Gordon didn't quite sound fully Oxbridge, however. I would later tell him that he sounded like a Scouse Elmer Fudd. Andrew, I would soon learn, was by far the very best piper in the band. He had fantastic fingers and had been playing since he was a small child, as did all of Tommy's kids. At one time Tommy ran a family band called the Graham Highlanders. Joe called them the Von Trapp family of Liverpool. Andrew apparently had problems as a wayward teenager and had been estranged from his father for a while. Tommy got him back to playing the pipes and made him clean up his act. It seemed to be working. I didn't learn all of this during that first session, but I wanted to introduce the cast of characters.

Tommy explained that they were preparing for a band competition the coming weekend in Blackpool, but he'd love to have me play with them on other pieces we might know in common. He suggested I go downstairs to the dressing rooms to "blow in" my pipes. This warming up is necessary before one can tune them. So I went downstairs to do so, and they went back to rehearsing. At least that's what I thought at the time. John later told me they'd stopped playing once they were sure I was lost in my own sounds and then crept up to the closed door to listen to me. They wanted to find out what kind of player I was. After a bit, Tommy said to those crammed against the door, "Not bad, that." From him, it was high praise, and I was pleased to learn that story.

When I emerged from warming up downstairs, they were just ending their break, and Tommy invited me into the pipers' circle. We adjusted chanters to match pitch, tuned drones to match chanters, and exchanged suggestions for tunes in common. It all went well enough, and by the end of the evening Tommy asked me to join the band. It was an invitation I was honored to accept and couldn't refuse. That band became my family for the next three months.

14 | Surface Reflectance

My social life was launched, and my professional life was finally underway. Peppered moths were in the trap, both the black and white phenotypes. I was struck by how much lighter in shade British typicals were from their American counterparts. These looked just like the textbook photographs. The black ones (*carbonaria*) were indistinguishable from their American counterparts (*swettaria*). Both of these form names, invented by lepidopterists, refer to the same phenotype: the fully melanic form. In the years ahead I would show, through Mendelian hybridization crosses, that they result from alleles at the same gene locus (position on a chromosome).

The intermediates (*insularia*) were absent in the first few days and remained at low frequency through the entire summer. The agreement was that I'd save each *insularia* for confirmation by Cyril, as subjectivity was involved in grading them and I was then a newcomer to the field. They ranged in appearance from one extreme to the other. Fortunately, they were rare at Caldy Common and Cyril's catches were large, so mistakes in scoring *insularia* could scarcely affect the percentages recorded for what was essentially a striking, discontinuous dimorphism between the typical and fully melanic forms. Cyril was not at all bothered about grading the rare intermediates, although his colleagues elsewhere in Britain were. He recorded just three categories—typicals, *carbonaria*, and *insularia*—explaining, "Once you get into assigning categories [Kettlewell recognized five] of intermediates, you'll never get out. I just look at them and decide what they are." After Cyril checked my scoring a few times, he had confidence in my eye and turned the moth trap and the categorization of its contents over to me.

I noticed that his trap, unlike Dave West's, did not have a rain shield to protect the mercury vapor (MV) bulb, and I mentioned this to Cyril. He wouldn't hear of using one. He worried that putting a glass beaker or Pyrex bowl over the bulb would reduce the brilliance or alter the wavelength of the light and affect the catch. You can't argue with success. He had the largest continuous record of change in the frequencies of peppered moth phenotypes on the planet. Who was I to fiddle with that? Unlike people who did use a rain shield and the same MV bulb for years on end, Cyril went through a large number of bulbs each season, and at several quid a pop, it bit into his budget. Still, that didn't convince him to use a rain shield. On one rainy morning I arrived to find a dead bulb, cracked and with water halfway up the inside. The trap was nearly empty of moths of any kind, indicating that the bulb blew out early on. When Cyril asked me what I'd caught that night, I told him, somewhat exasperated, "Nothing. Zero. Nada." I reported that the bulb was broken and cautiously suggested that a rain shield would have prevented that. "Yes," he said. He'd think about it.

He thought about it but stuck to his guns about not changing how he'd done things for 25 years. A few bulbs later, after similar conversations, I timidly asked if a reduced catch, perhaps caused by using a rain shield, was worse than losing an entire night's data when the bulb blew and no moths came to the trap? "Yes" he drawled. "I need to think about it." We probably went through nine bulbs that summer. For comparison, Dave West had used the same bulb for about nine years by that time.

I put the first several moths, with both phenotypes represented but without collars, into a background test pen (my version of Kettlewell's barrel). I could barely wait until the next morning to see where they would be. From Kettlewell's previous work, I fully expected to see a bias, with melanics choosing black stripes and typicals choosing white stripes. Having that result wouldn't confirm Kettlewell's contrast/conflict mechanism for background selection. Nor would it weigh in on the side of Sargent's coadaptation argument. What it would show is that these moths, dimorphic for body color, were also dimorphic in their behavior. That would have been mind blowing to see. It would then be up to me to determine how they made their choices.

Much to my disappointment, my mind wasn't blown. There was no striking difference in the backgrounds chosen by the black and white moths. Perhaps the distinction between them was subtler, and I'd have to run many more moths through the tests to see it. Perhaps I needed a much better apparatus to bring out behavioral disparities between the phenotypes, assuming that they existed. I wasn't discouraged to the point of giving up, but I knew then that solving this problem was going to take more work than I'd hoped.

I decided to pay closer attention to the light reflectance from the moths' wings and the background surfaces I was offering them. Sargent had done this, but I hadn't yet put in the effort. Without the right gear, measuring reflectance from a piece of fabric is a lot easier than doing so for a moth's wing. I tried spreading out several wings from multiple moths, as Sargent had described, to increase the reflectance target size; used photographic light meters; and compared the readings with a barium sulfate standard. My crude methods were later refined by a colleague, Rory Howlett, who appears later in this story. His far more sophisticated gear employed a Macam Photometrics SMU 101 spot-measuring unit. The point here is that while *carbonaria* look uniformly black to the human eye, the typicals are not actually white. They are gray in overall appearance, but there is considerable variation in light reflectance across the surface of a *typica* moth's wing, caused by the speckled distribution of black and white scales. There is also variation among the wings from different *typica* moths. Rory attempted to replicate a generalized pattern by painting the backgrounds offered to captive moths as resting surfaces.

I hadn't yet met him, so I tried two separate approaches. One was in hope of locating a reasonably suitable fabric that resembled the mottled gray of a typical moth. Finding black in any texture was easy, but getting just the right pattern for *typica* proved to be problematical. Checks, blocks, splotches, gingham, tweed, and houndstooth designs on various fabrics, ranging from linen to felt, were scrutinized by me and the moths. After haunting the fabric stalls at a vast flea market in Birkenhead, I aroused the curiosity of a salesman. "What are you going to do with this material?" he asked.

I thought about just brushing him off with some general response, but he seemed genuinely curious. "I want to find out which of these patterns moths prefer to hide on during the daytime."

His eyebrows shot up. "Is someone paying you to do this?"

"Well, actually, yes," said I.

"Then I'm going with you." He stood up and started to remove his apron as if he were coming with me. We had a good laugh. He didn't even ask me if I was an American. My phony English accent must've been improving.

Whatever fabric I tried in the moth pen, I got fairly consistent results as long as its gray color had about half the reflectance of standard white.

Moths avoided white, but they did settle on gray and on black. I even tried gradients in the fabric's shade. What became clear was that the moths were not indifferent to the resting backgrounds offered to them in captivity. It also became apparent that the typicals and the melanics did not differ in their preferences.

I took me all summer to learn these things. I'm not complaining, though. I had a great time, the whole time.

15 | How to Pick Up a Moth

As I arrived for work one morning, Lady Clarke informed me that I had two phone messages from the previous day. One was from an automobile body shop about money I owed them for mending a car, and the other was from someone who plays bagpipes. She handed me the phone numbers. I couldn't tell from her expression whether she was concerned about the car or curious about a call from a bagpiper. I was worried about how much the body shop was going to soak me for, and I was puzzled by the call from a piper, because I hadn't given anyone in the Liverpool band a phone number. Ah yes, I suddenly remembered, my first night in England I'd left the phone number at a pub. "Thank you. I'll ring them. Sorry to trouble you with this."

Lady Clarke was far too polite to ask about the calls, but she looked expectant, as if I might say more. So I did. "I had a bit of a bump when I first bought my car—wrong side of the road. My fault. Nothing serious. Just scratched the paint on the other car." That didn't satisfy her curiosity, but she still said nothing, just kept looking at me inquiringly and smiling politely. "The other phone call," I speculated, "must be from the bagpiper who visits the little pub in Caldy Village. I asked the manager there to give him your phone number next time he dropped by." That didn't do it, either. Lady Clarke was still smiling, waiting, almost bursting to ask why on earth would I do that? I resolved the mystery by confessing, "You see, I play the bagpipe, and I was hoping to find a band I might join for the summer."

She didn't actually say "Yikes," but her eyebrows indicated that's what she was thinking. Just then Sir Cyril came out of the house, into the tight passageway between the back entrance and the garage where Lady Clarke and I were chatting. She said in the most pleasant tone, totally concealing her astonishment, "Cyril, did you know that Bruce is a bagpipist?"

"Yes," he drawled absently.

"You did? You knew that Bruce is a bagpipist?"

"No," he corrected himself. Turning to me, he asked, "Are you good?"

"Not all the time," I admitted. "Only when I'm in tune."

"I should like to hear you. You must play for us sometime—when you're in tune."

I don't recall playing for them that summer, but I did give a brief garden recital to his assistants one afternoon when they gathered to celebrate a birthday. I hope I was in tune.

After our chitchat, I went into the garage lab and put collars on a few of the moths I'd caught that morning. After they'd recovered from the anesthesia, I took a specimen into the house to show the Clarkes. During daylight hours, peppered moths remain dead still once they've settled, so one may handle them gently without alarming them. I carried the moth on a small slip of paper and handed it to Sir Cyril. It was a black moth sporting a white collar. He held it in his hand and smiled broadly. "How canny," he repeated several times.

My purpose in showing him the collared moth was twofold. First, I confess, was pride in my idea, and I wanted to show it off. Who on the planet knew more about peppered moths than he did? I wanted his reaction. Second, and not so vainglorious, I wanted to impress on him the need for healthy, undamaged moths for my experiments. His method of trapping differed slightly from mine, and I hoped he'd permit a slight tweak. As I'd learned from the rain shield suggestion, Sir Cyril was resistant to change. He had his reasons for that. This alteration was much subtler. The trap, the bulb, and the funnel would stay exactly the same, but I wanted to substitute flat pieces of cardboard for the egg cartons he used inside the trap.

Once caught, a moth bats around until it finally settles on a surface and gives in to the idea that it's daytime. There it usually remains until removed from the trap in the morning. Egg cartons placed inside the trap provide a lot of surface area for moths to cling to, and the individual egg compartments help protect them from further disturbance by other moths banging around during the night. Made to order! But the problem with egg cartons is that it's sometimes hard to remove a moth that has nestled deep into a compartment, head first. One has to pry them out very gently, to prevent damaging them in the process. That effect had never been a problem for Cyril. When I asked him what he did with the moths he caught, he said, "I nip them."

Nipping was sound practice for a population survey, as there was no risk of recapturing the same moth once it was dead. Riki, however, had a soft spot in her heart for all living creatures. So do I, and I understood her feelings completely. She couldn't bring herself to "nip" them, so she marked and released them well away from the trap. The mark served to let her know that the moth, should it return, had been trapped previously, so it wouldn't be counted a second time. One mustn't claim to have counted 100 moths by counting the same 10 individuals 10 times! After marking the moths, Riki would drive off with them somewhere, far enough away so recapturing them was a rare event. Unlike her boss, Riki was very much interested in removing moths from the egg carton without injuring them.

I showed her the method I used. I'd take a small piece of paper and slip it gently under the moth's front end. Done correctly, the moth steps onto the paper, allowing me to pick it up and transfer it to a cage or just about anywhere else I might wish to relocate it. The moth remains undisturbed and unharmed. Doesn't lose a scale. When I demonstrated this to Riki one morning, I reached into my pocket for a piece a paper but didn't have one, so I folded up a £10 note and removed the moth using the money. When I'd finished the demonstration, Riki asked in mock seriousness, "Must it be a *ten*-pound note, or will a fiver do?"

More important than the denomination of the bill is the direction in which it's presented to the moth. A moth will cooperate beautifully if the paper is slid under it from its front (anterior) end. Sliding the paper in from its rear (posterior) seldom works and can agitate the moth. For this reason, I prefer to provide flat cardboard as a resting surface, so I can easily approach the head of any moth settled on it. The problem with an egg carton is that a moth frequently rests with its head tucked deep inside an individual compartment, making it difficult to slip the paper under the moth from the anterior direction. Too much fiddling around may stimulate it to bolt. I didn't learn this detail from books. There is no substitute for hands-on experience to understand what Barbara McClintock meant by the necessity of gaining a "feeling for the organism" to learn its secrets.

Cyril, of course, had no trouble just putting his finger into an egg compartment and prying out the moth, but moths removed in this manner were often mangled by such rough handling. He and I had interests in different secrets. Cyril wanted to know how many of what kind (melanic vs. typical) were flying about, and I wanted to learn how living, healthy moths might select daytime hiding places. Thus I required *live, healthy* moths from his traps. He didn't care at all about their condition, as long as he knew how many of what kind came in and that none had gotten away to be re-caught and recounted. When I asked him if he'd permit me to substitute flat cardboard for the egg cartons, I fully expected him to say no, or that he'd need to think about it. I was pleasantly surprised when he said, "Use whatever method you wish. It shouldn't make any difference, as long as they don't escape." I assured him that I hadn't lost one yet, but there were no guarantees the odd one might not get away. "Yes," he said.

I withdrew upon hearing his protracted "Yes," as it signaled that he was rethinking his answer, and I worried he was about to change his mind. I'd have to learn not to be so candid. No guarantees, indeed. Why did I need to say that? I'd had the permission I'd sought, but then I blew it by saying too much. I'd given him something to worry about, and when it came to altering his tried-and-true methods, an action that might jeopardize in any way the accuracy of his records, he'd worry a lot. I am happy to report that he didn't rescind his permission for me to remove moths from the trap using flat cardboard and slips of paper, and I didn't allow a single moth to escape that summer. I can't make that claim for every season. The odd one does get away from time to time, but these are very rare events that can be counted on the fingers of one hand after more than 25 years of trapping.

16 | Birch Moths

Seasonal fluctuations in population densities are familiar occurrences in lepidopteran species. Peppered moths are not exceptional in this regard. Our analysis, published in 1996, of Cyril Clarke's trapping records over a 36-year period indicated that the wide fluctuations are cyclic. Data from other workers are also suggestive of cycles, but their data sets cover much shorter periods, and trapping efforts from year to year vary greatly. Assuming that these are consistent, catch records should reflect local abundance. Simply put, there are good years and there are bad years. Whatever ecological factors are responsible remains unclear in most instances, and this is certainly the case for peppered moths. Summer weather conditions; the severity of the previous winter; shifts in predators and parasites; diseases; resource competition with other species; and, now, the use of pesticides are all suspected of playing roles in population fluctuation. For peppered moths, little attention has been paid to this problem. Cyril Clarke's season totals ranged from just over a hundred to over a thousand, indicating a 10-fold fluctuation in population size.

The year before I went to work at 43 Caldy Road, he'd caught 689. The following year, he captured 860. My first year there, in 1984, we only trapped 352. While not a big catch by Cyril's standards, it exceeded what I'd caught at Mountain Lake by over 200. More importantly, of the 352 peppered moths we caught that summer, 61% (215 of them) were black (*carbonaria*). Of the remaining 39%, 120 were pale (*typica*) and 17 were sooty intermediates (*insularia*). I was awash in the material I needed to reexamine background selection in this species!

I had enough material to try several pilot studies, hoping to find an approach that would provide unambiguous answers. No matter what you do, you'll get a result. But is there more than one possible explanation for that result? The work isn't over until all possible answers have been ruled out, except one. (Yogi Berra probably said this.) We arrive at our conclusions by a systematic process of elimination, not by settling for the first answer we find appealing. To illustrate, suppose I put several black and pale peppered moths into a test pen that has only black and gray panels as the options for their daytime resting places. Then I score where each moth is sitting the next morning. Suppose all of the black moths are on the black panels, and all of the pale moths are on the gray panels. I might decide from those results that a difference in the behavior of black and pale moths exists. But perhaps I'd be premature in that conclusion, especially if only a few of each moth phenotype were used in the test. So I'd have to do the test again, to ensure that the first result wasn't a fluke, and it'd be a good idea to do that repetition with a different set of moths, on the odd chance that those particular specimens were atypical. If I repeated the experiment again and again and kept getting the same result, I could, with increasing confidence, conclude that there is a real difference in the behavior between the black and pale moths. This would be a *qualitative result*, meaning there is no overlap between the behavioral phenotypes. Black moths settle on black backgrounds, and pale moths settle on gray backgrounds. Period. This conclusion is unambiguous.

But such a conclusion, however firm, wouldn't in the least resolve the dispute over Kettlewell's contrast/conflict (pleiotropy) model versus Sargent's genetic (coadaptation) model about the mechanisms by which moths make background choices. It would demonstrate only that the moths *do* make different choices, not *how* they make them. The problem for me was to initially confirm that the several peppered moth phenotypes actually make different background choices. I was having trouble accomplishing that very first step. Through their test scores, the moths seemed to be telling me that I was on a snipe hunt, trying to discover a mechanism to explain differences in their behavior that didn't exist. But I was not yet confident enough in my results to argue that Kettlewell was wrong. He reported a behavioral polymorphism that I couldn't confirm, try as I might.

No sense worrying about contrast/conflict at this point. Kettlewell never did an experiment to test that idea in the first place. It was pure speculation on his part, which was promoted—without supporting evidence—by his globally influential sponsor, E. B. Ford, and consequently assimilated into the lore and literature by uncritical enthusiasts. First things first, I reminded myself. Before examining the explanation for a phenomenon, I had to determine if there indeed was a phenomenon to explain. At that point in time,

I desperately wanted to find out if peppered moths that were polymorphic for body color phenotypes actually were also behaviorally polymorphic for background selection. Kettlewell reported that they were, and he published data to back it up. I had no such data yet, but I now had the moths to find out if Kettlewell's claims were true. I sure hoped he was right. I didn't give a hoot about contrast/conflict, but I wanted the behavioral polymorphism to be real. That'd give me something fun to explore.

The qualitative result in my above supposition was the presentation of an extreme on a continuum of possibilities. Rarely is the scoring for any phenotype that uncomplicated, although it is generally much easier to measure a physical character, or trait, than a behavioral one. For example, I could look at a moth and decide in an instant if it's a black moth or a typical. I might have a bit of a problem with some shade gradations for *insularia*, but never, absolutely never (as far as I know) would I misidentify a *carbonaria* as a *typica*, or vice versa. (I might carelessly make an incorrect entry in a data book, but anybody can tell these phenotypes apart at a glance.) It's a fixed, physical phenotype. It doesn't change for any given moth from day to day. Usually, distinctions are not so clear for behavioral phenotypes.

To hammer this point home, consider the determination of A, B, and O blood groups in humans. My blood type is determined by my genes. It hasn't changed since I was born. I'm still the same blood type in my dotage. In this instance, the phenotype is a reliable indicator of the genotype. My blood pressure, on the other hand, has changed dramatically over the course of my life, and it alters with the time of day. While there may be a heritable component, blood pressure is not a reliable indicator of genotype, as it's influenced by numerous environmental factors. Similarly, genetic analysis has demonstrated that the *carbonaria* phenotype of peppered moths is a reliable indicator of genotype. They are black because they each carry at least one copy of a gene that results in the overproduction or deposition of the pigment melanin, whereas pale forms are homozygous (having identical alleles for a particular trait) for the recessive *typica* allele. But what about background selection behavior? Surely that's not so easy to measure, as my own experiments were showing me. Could it be that there were variances in the behaviors between melanics and typicals, but I wasn't putting the moths into the right environments to accentuate those differences sufficiently to observe them? A test pen is an artificial environment (a cage), and no doubt captive moths behave abnormally in many respects. They still do some things normally: they fly and walk around inside the cage at night and remain inactive (clamped down) during the day. Is there at least some of their usual behavior involved when they finally settle on a spot (either a light or dark surface) to clamp down?

Kettlewell used captive moths held in a cider barrel, and he did report that the melanic and typical phenotypes behaved differently from one another with respect to background selection. But his data did not show invariant responses on the part of melanic and typical moths. Some melanics rested on the white surface, and some typicals rested on the black background. What he showed were statistically significant tendencies, with melanic moths settling on black surfaces twice as often as on white, and typical moths settling on white twice as often as on black. (I reviewed his numerical results in Chapter 5 and will not repeat them here.) What is essential is that there may be circumstances where these tendencies are either more pronounced (a qualitative result, with no overlaps, representing one extreme) or less pronounced (to the point where no tendencies, or differences in tendencies, are apparent). On a possible continuum of degrees of difference, perhaps my experimental conditions were at the wrong end of the spectrum. After all, I wasn't repeating Kettlewell's experiments in exact detail. I wasn't employing an actual barrel, with black and white paper pinned inside. Perhaps crucially, I wasn't using the same source of moths. (More on this point later.) But I hadn't abandoned hope, nor was I prepared to challenge Kettlewell's results just yet. I was still only a beginner and had much to learn.

But some things I already understood. Observing no difference in the behavior of melanic and typical moths allowed for two possibilities: (1) either no differences in their behaviors existed, or (2) differences were present, but I hadn't devised a suitable experiment to reveal them. This is an old chestnut in science. You can sometimes convincingly show the existence of something, but you can never unambiguously demonstrate the nonexistence of something. At best, you can move from an absence of evidence to evidence of an absence by repeating and modifying experiments. Success (evidence of absence) comes from repeated failures to confirm earlier work. Life would have been so much more fun for me had I obtained the same results as Kettlewell had. But I hadn't.

Still, it is very important to point out that the moths were not indifferent to the resting backgrounds available to them in captivity. Taken as a group, they showed a statistically significant bias for backgrounds with different surface reflectances (generally preferring black to white, or gray to black). But no clear, repeatable behavioral distinction was evident between moths of different physical phenotypes (*carbonaria* and typicals) with respect to their background preferences. What I did learn, however, was that individual differences in background selection existed. It took me two more years to nail this down and offer a sensible explanation for why my data contradicted Kettlewell's.

For the remainder of my first summer as a guest researcher at the University of Liverpool, I commuted twice daily, seven days a week, from my garden apartment in Birkenhead, across the width of the Wirral Peninsula, to the Clarkes' home in West Kirby at 43 Caldy Road. Most of my time there I spent alone, arriving before his assistants were due and usually leaving with the day's catch shortly after they came in. There was always time for a friendly word or two, but we all had our work to do. On weekends the Clarkes would be home, and we chatted briefly as well, mostly starting with Cyril's greeting from his lavatory window—"How many?"—and having coffee afterward. Sometimes this led to an invitation to join them later on, when this or that person was to pay them a call. Lady Clarke would say, "Please do come entertain us!" This would be reinforced by Cyril's inevitable "Yes."

On one such morning, I lamented that I was having trouble coaxing the moths into selecting matching backgrounds. I then went on to say that perhaps I should try a more natural background, such as real tree bark covered in soot versus bark encrusted with lichens. Kettlewell had also done that as a follow-up to his first barrel experiments. "But there are no lichens around here," I complained. "Where are the lichens?"

I learned from discussions with his assistants that Cyril had unsuccessfully attempted to transplant lichens into his garden, and he was still at it. Riki would bring them back with her from rural Wales.

"It's odd, don't you think," I continued, "that the typicals should be making a comeback here in the absence of lichens?" I then went on to argue that if typicals are indeed protected from detection by birds by these moths' resemblance to lichens, then we should predict that the recovery of lichens in this region should precede an increase in the frequency of moth phenotypes that depend on them for camouflage. "How can the hiders return *before* the return of their hiding places?" I demanded.

Cyril replied to that with his drawn out "Yes," this time meaning that my point was well taken. He then added that however long it might take for lichens to reappear in response to a reduction in atmospheric sulfur dioxide (SO_2), there was already a gradual but conspicuous lightening of the whole region following the Clean Air Acts. Surface reflectance had changed dramatically throughout the area, and every adult who grew up there was aware of this. He described the grime and sooty conditions that had existed well within recent memory, and he told a story about how people determined in which direction the wind was blowing when a little white dog went outside. Within minutes it would become blackened on the windward side, while the lee stayed white. Cyril was full of colorful anecdotes to illustrate any point.

Roy Leverton put it vividly in his beautiful book, *Enjoying Moths*: "Those who did not experience it may not appreciate the severity of atmospheric pollution in industrial and urban areas before measures were taken to reduce it. For the first 18 years of my life I lived in Salford, part of the Manchester conurbation. Only after moving to Sussex in 1964 did I realize that black is not the natural color of nasal mucus" (p. 50).

The rest of Leverton's personal recollections are well worth reading, particularly his amazement at how much cleaner everything was in Salford, including the atmosphere, when he returned some 30 years later.

Perhaps the general lightening of habitats did not afford typicals the full protection their phenotype might enjoy in lichen-encrusted woodlands, but they were faring better than *carbonaria* under the soot abatement, judging from the reversals in the frequencies of these phenotypes underway on the Wirral.

The government-mandated effort to reduce atmospheric pollution was paying dividends. But smokeless zones in Britain weren't created for the benefit of peppered moths. The regulations were imposed because human health was adversely affected by air pollution. Cyril seldom wrote or spoke about industrial melanism in peppered moths without mentioning that bronchitis is exacerbated by breathing smoke and SO_2.

We now live in an era of global climate change and are reminded daily of increasing levels of other gases in our atmosphere, chief among them CO_2, referred to as one of the greenhouse gases. Not everyone has been inside greenhouses, but most people in the modern world have experience with cars and know that a vehicle parked in the sun, with its windows closed, gets deadly hot inside. The visible light from the sun passes through the glass windows and heats up the interior—the seats, the dashboard, the floor, the steering wheel, and the air itself. These objects send back some of that solar energy in the form of infrared radiation, which doesn't readily escape through the window glass. The heat is trapped inside the car until someone opens the windows. A vehicle parked in the sun is a microcosm of our planet because certain gases, such as CO_2 and methane, reflect back infrared rays coming up from Earth's surface and, in that way, act like car windows. To keep the planet from overheating, our current goal is to reduce the levels of these gases. While we have been remarkably successful in employing various "scrubbers" to reduce a wide range of noxious compounds emanating from factory smokestacks, and by using cleaner forms of fuel, why can't we just as effectively reduce greenhouse gases? If we can reduce SO_2, why not CO_2? It's a question of scale. CO_2 is not an impurity in the fuel. Rather, it is the direct product of combustion. The more carbon compounds (e.g., coal, wood, petroleum products) we burn (fire is a form of rapid oxidation), the more CO_2 is released into the atmosphere. It's a massive problem. "Clean" coal is not the solution to global warming. No carbon-based fuel is.

As I wandered around the Clarkes' garden each morning, with frequent walks along Wirral Way and the Dee, and across Caldy Common, it occurred to me that the place was dripping with silver birch trees (*Betula pendula*). *Betula* is the name of the genus to which various birch species belong. Peppered moths have been reassigned to several different genera by taxonomists before their current placement in *Biston*, but the species name, given by none other than Linnaeus, has stuck: *betularia*. Eureka! I wondered if maybe Linnaeus knew more about this moth than we do. All this time we've been calling his "birch moth" a "peppered moth." Hmm?

During one of my morning chats over coffee, I asked Cyril's assistants if they knew where I might get some birch logs to use in building a moth cage. Without hesitation they directed me to nearby Royden Park, where I was to meet Ranger Andrew Brookbank. Once there, I introduced myself to the ranger and explained my requirements to him. Before I had time to protest what he was about to do, he'd donned his hardhat and goggles, started his chainsaw, and cut down a tall, slender birch tree for me to haul away. He very kindly sawed it into convenient sections to fit into my Hillman Hunter. In public parks there are strict rules about gathering and removing wood, and certainly cutting down trees is strictly forbidden (without a permit). Ranger Brookbank explained that the section we were standing in was being managed, and the birch trees were to be cleared. His chopping one up for me was just part of the removal program. Indeed, he assured me, I could have as many as I wanted from that part of the park, and he'd gladly cut them down for me.

Back at the ranger station, he showed me an aerial map of the region and explained that clearing out the birches was part of an ongoing program to restore managed heathlands that had been overgrown by the encroachment of these trees, which had expanded into the region following the Clean Air Acts, when this area was declared a smokeless zone. That meant people could no longer use fireplaces to heat their homes, so burning coal and wood for that purpose abruptly stopped. Because coal costs money to buy, and birches grow freely, people of limited financial resources had previously kept this tree species well under control. As wood goes, it wasn't in great demand, except for heating homes. Once that practice stopped, the birch stands expanded rapidly. I was riveted by his words.

Birch, as a "weed" species, wasn't just a local problem on the Wirral. Over the next few years, I acquired published references describing similar circumstances elsewhere in Britain, including one titled "Losses of lowland heath through succession at four sites in Breckland, East Anglia, England," published in *Biological Conservation* by R. H. Marrs et al. in 1986. The recovery of the typical phenotype of peppered moths seemed to be occurring in concert with the expansion of birch trees following the Clean Air Acts. Maybe, I considered, lichens weren't crucial to the effective camouflage of this phenotype.

I wasn't the first person to come up with that idea. Kettlewell himself suggested it in a paper he published in *Heredity* in 1955: "That the wild *B. betularia* population of Rubery [a polluted woodland devoid of lichens] contained as much as 10.14 per cent. *typicals* is, in my opinion, due to the advantage of this form in the surrounding birch woods of which there were many in the neighborhood (pp. 339–340)."

With the retelling and abridgement of the peppered moth story, the importance of lichens was kept and birches were forgotten, so much so that whenever I bring this up, it's as though I'm shouting that the emperor is wearing no clothes! Some even consider the lichens-versus-birches issue a threat to the very fabric of industrial melanism as an example of natural selection. That is the sheerest nonsense! Where lichens exist, they may aid the effectiveness of the camouflage for typical peppered moths, but clearly lichens are not crucial to these moths' survival. Indeed, typicals seem to thrive rather well in birch stands, with or without lichens.

"We have a mating!" Lady Clarke greeted me as I arrived for work one morning. The previous day we had caught a female *typica* in the trap. Catching a female is a rare event, as males of this species do most of the flying, although the females are powerful fliers, unlike some moth species in which females don't take to the air at all. Moreover, in some species, females don't even have wings. In peppered moths, the females fly when they first emerge. After her initial dispersal flight, a female alights, takes up a characteristic posture (wings up, abdomen extended), and "calls" males to her by releasing pheromone. Once pairing occurs, a mated couple remains joined end-to-end throughout the next day. By dusk the pair separates and the male is then free to fly about looking for a new mate. The female seldom takes wing again after mating, instead devoting herself to laying eggs. In the process, she provisions the eggs by reallocating resources from her own tissues, so, by the time she's finished depositing them, her muscles atrophy to the point where she can't fly. Thus most of the moths caught in a trap, at least for this species (and many others), are male. Catching a female is cause for celebration. Doubly so, because usually she's a virgin. This provides an opportunity to make hybrid crosses for progeny testing and for a ready source of females the next year.

Lady Clarke's jubilant exclamation that "We have a mating!" was understandable, because it meant we'd soon have eggs. If they were fertile, we'd shortly have caterpillars to feed, and then pupae to store until next year. This was good news for me, because the mating was between a *typica* female and a *carbonaria* male. We'd put in three *carbonaria* males to be sure one of them would be successful, and I wanted to have both alleles to take home to the States, in order to keep me going with moths from the United Kingdom.

As we stared like voyeurs at the mating pair—the end-to-end, delta-shaped moths forming a diamond shape as they clung to the net of the hanging cage—Cyril said, with a wry smile, "When a *carbonaria* and a typical mate, one of them places both at risk."

I had heard this before. Kettlewell said it. The idea was inspired by his notion that black moths normally settle on black backgrounds and pale moths normally settle on pale backgrounds, but if a mating occurred between a black moth and a pale moth, then one or the other of them would be conspicuous to a predator by day, no matter which colored background they were on. Thus both moths would potentially be at risk. Furthermore, some suggested, such selection against mating between unlike phenotypes could lead to assortative (nonrandom) mating. By this time, with birch trees on my mind and no evidence that moth phenotypes showed background-matching polymorphism, I was ready to challenge this speculation. I went into the garage lab and fetched a hunk of birch log, splotched with black limb scars across mottled gray bark. I held it up behind the hanging cage, so the mating pair clinging to the net was posed against the backdrop of birch bark. I asked Cyril, in response to his claim that in a *typica-carbonaria* match, one of the moths places both at risk, "Which one of these do you think is the conspicuous moth on this background?"

He looked at the display as if he was searching for the right answer, but all he said was "Yes." Then he went into the house.

17 | Cultural Assimilation

The pipe band held two formal practices a week during competition season. One night was devoted to practice chanter work, and the other was a full band practice on pipes. Weekends were busy times, involving travel to competitions elsewhere in the United Kingdom or playing in local parades (or carnivals) for honoraria. Sometimes we did both on the same weekend. Tommy used the parades to issue us new reeds, so they'd be well blown in before upcoming contests. When nothing was on the books, he called everyone to Sunday morning band practices in car parks. Most of this is familiar stuff to people in pipe bands. What was a new activity for me was busking. I didn't like it at all, but I went a few times for the experience. Others in the band did it regularly. It was their chief source of income.

Here's how it worked. A few nights a week, several pipers would meet Tommy at the recreation center in his apartment complex, where they'd be groomed and tuned and sent off as teams to work the pubs of Liverpool for money, playing a few tunes in one and then moving on to the next on the assigned circuit for that evening. It loosely reminded me of Charles Dickens's school for pickpockets in *Oliver Twist*, with the part of a more lovable Fagin being played by Tommy. There was certainly no stealing going on with Tommy's buskers, but it was a much better organized enterprise than it appeared to the unsuspecting bar patrons who happily chipped in a few bob when the figurative hat was passed among them. Instead of using an actual hat, one of the players would circulate around the bar and then go from table to table, jiggling a tall can with a slot on top into which money (both coins and bills) was shoved. He encouraged donations by chanting something about needing to buy band uniforms for the upcoming lads and lassies. It was staged as a fund drive for the City of Liverpool Pipe Band, not a ragtag gaggle of buskers hoping to supplement their personal incomes. Most of the money initially did go toward buying uniforms and gear (sizable expenditures, to be sure), but now the "surplus" was counted out at the end of the evening, with equal shares going to each member of the team, minus the cut reserved for their mentor Tommy.

After several invitations to join a busking team, I reluctantly decided to have a go. My group included Tommy's son Andrew, as well as Joe, the chemistry grad student. The division of labor was that Andrew and I would do the piping, and Joe would do the talking. This was good casting. Andrew was the best piper in the band, and we played well together, but he seldom spoke. I didn't have the proper local accent. If the drinking blokes in a Liverpool pub learned they were being hustled by an American professor, they'd surely tear him limb from limb. So, except for my pipes, I was mute. Joe had an easy gift of gab in rapid-fire Scouse and loved to mingle with his people. Shyness was not part of his vocabulary.

The routine was well rehearsed. We'd drive in my car to within a short walk from a busy pub. Joe would first go into the pub alone, dressed smartly in his band uniform (Glengarry hat, jacket, tie, sporran, hose, kilt—the full nine yards), and ask to speak to the manager. He'd inquire if it would be all right to have a few pipers come inside to play a few tunes to solicit contributions in support of the local pipe band. Rarely was he denied. The patrons nearly always seemed delighted to hear us play, and they generously stuffed money into our collection can. Indeed, some people wanted us to stay longer and keep on playing, and they would offer to buy us drinks. Fast-talking Joe would explain that we couldn't drink at this point in the evening, because we had to make our rounds to other pubs. The pub's patrons were aware that the law against drinking and driving in the United Kingdom was strictly enforced, and we knew drinking and piping didn't work well at all. Only after busking in our final pub, before closing time, were we willing to knock off for the evening and quaff a few brews, except for the designated driver.

I recall the very first time I went busking, when an enthusiastic patron of the arts had set up three shots of whisky on the bar at our first stop. I knew this would be a bad start for the evening if it kept up, so I attempted to refuse the drink. But how could I do this gracefully, without speaking? Joe couldn't get us out of the situation, so it was gulp and "Ta" and out the door. Other close calls came when inebriated blokes would shout into my ear while I was playing. This was how they made requests. The trouble was, I couldn't understand what they were saying. I usually said nothing and just kept playing, hoping Joe or Andrew would come to my rescue. They never did, and it soon became clear to me that they rather enjoyed my anxiety. On more than one occasion while I was playing solo, I would hear someone in the crowd shout out, "He's a Yank!" I was unnerved the first time this happened, wondering how they could possibly know that. My piping didn't have an American accent, at least one of which I was aware, and I hadn't spoken a word. Then I realized the shouts were coming from Andrew (the silent one). He did this for amusement. Scouse humor.

Toward the end of summer, the band traveled to Glasgow for the World Pipe Band Championships (or, to those of us in the game, simply "the Worlds"). They really are the Worlds (unlike American baseball's World Series), as bands from all over the entire globe came to Glasgow to compete, including places from very far away, like the United States, Canada, Europe, Africa, Asia, Australia, and New Zealand (where there are more pipe bands per capita than in Scotland).

Along the motorway en route to the Worlds, traffic ground to a complete stop across all northbound lanes. The jam extended for as far as we could see. To most people, traffic snarls are annoying at best, but to the City of Liverpool Pipe Band, this one was a Scouse opportunity to do a little busking. A captive audience! Almost instantly, our group of pipers were strolling among the trapped cars, followed by the can bearer. I was far too nervous to join them, worrying that at any minute the traffic would start to move and we'd be caught on foot in the middle of vehicles eager to make up for lost time. My daughters had joined me by this point in the summer, and I tried unsuccessfully to persuade the older one not to dance the fling on the motorway. Thankfully, they all piled safely back into the cars when it became obvious that traffic was about to move. I was quite surprised that they actually made some money. I'd have guessed people would've just rolled up the windows and ignored them, but it was a warm, sunny day, and stalled vehicles get too hot inside to keep their windows shut against determined buskers outside.

We couldn't field a band to compete at the Worlds, because too few drummers made the trip to Glasgow. So we went busking the night before, more for fun than for money. Some of the native Scots weren't amused. I overheard a bevy of crones at a table complain, "They're English! What cheek!" But after a fairly pleasant evening, at least from our perspective,

Andrew corrected a local yob who thought I was also English. "No. He's a Yank," said Andrew, thoroughly enjoying himself. "Yer no," said the incredulous Scot. I confessed that I was indeed an American, at this point preferring not to be thought of as English among hostiles. Then the Scot challenged me to "Say somethin' American!" To this I recited as clearly and flatly as I could manage, "The rain in Spain stays mainly in the plain," in an earnest attempt to convince him of the truth. He scoffed at my effort. "That's the most rubbishy American accent I ever heard. Yer English!" When he cried out to his mates to "gi' a listen," I figured it was time to call it a night.

Busking was an activity I never enjoyed, but I easily understood why "me mates" in the band did it at every opportunity. They made a lot of money! On our first night out, back in early June, between stops in each pub we'd empty the collection can of all but a few coins. We stored the take in a large bucket locked in the boot of my car. As the evening wore on, the bucket filled up—and I mean *filled up*—mostly with coins, but also with a fair amount of paper money. When we counted it out and divided it into stacks, I exclaimed, "Geez, a guy could make a living doing this!"

"We do," said Andrew. "Don't you go busking in America?"

"Well, some people do, but I've never done it. Not my cup of tea. Not my bag." I didn't pay close attention to counting out the money, and I'm sure they shorted me, but I still made out well enough on those few occasions when I agreed to go with them.

While the Liverpool pubs made me nervous, a potentially life-threatening outing occurred in Nottingham. The band had traveled there to compete, and on the return trip, the contingent that was crammed into a rented minibus stopped at a pub for a pint. Most of us were pipers, but we had gear as well, including the bass drum that, earlier in the day, had dropped onto the road when the back doors popped open. Looking in the rearview mirror, I saw it disappear as it rolled down a long grade into heavy traffic. Fortunately we retrieved it undamaged, promising a good day. In the car park at the pub, we emerged from the minibus like circus clowns armed with bagpipes.

Nottingham at that time was at the center of British national news, because coal miners were on strike as their powerful union leader, Arthur Scargill, was locked in a heated battle with Prime Minister Margaret Thatcher about what was to become of the mines. Collieries around the country were being permanently closed under government-directed reorganization, and this meant massive layoffs were in the offing. Whether on strike or laid off, miners were desperate and clashed with the police daily. As the strike wore on, union sympathizers raised money to support the hard-pressed miners' families. As we sat in the Nottingham pub enjoying beer and conversation, a crew of union sympathizers came in, asking for donations. They had their own collection can. I felt obliged to drop in a few coins as they shook the can vigorously under my nose. They thanked me for my generosity by pressing onto my shirt a bright yellow plastic sticker with their campaign slogan, "Dig Deep for the Miners," in coal-black lettering. The sticker showed that I had done my part and encouraged others to do theirs. By the time the union sympathizers had made their rounds and left the pub, all of us in the band were sporting the bright yellow "Dig Deep" stickers on our shirts.

We were still in kilts, so it was clear to the patrons that we were an off-duty pipe band. After buying our own first round of beer, some bar patrons asked, "How about a tune, pipers?" It didn't take much coaxing, because this meant (1) more beer was on the way, and (2) it would be free for as long as we played. The request also presented us with an opportunity to do some subtle busking—no collection can, but we'd leave our pipe cases open to invite the odd coin.

We stood in a circle and played away, and money came pouring in. They really seem to love us here, I thought. Then I discovered why. "Where's my sticker?" someone cried. A belligerent bloke was demanding that I give him a "Dig Deep" sticker, explaining that he'd just put a donation in my pipe case. It didn't take me long to figure out that as new patrons entered the crowded bar, they assumed from the "Dig Deep" stickers band members were wearing that we were piping to raise money to support the striking coal miners. If they came to the realization that we were not affiliated with that noble cause, there was no telling what might happen. I imagined that they would regard us as contemptible imposters and pound us to death with our own drones. What to do?

Without saying a word to the chap demanding a "Dig Deep" sticker, I peeled the one from my own chest and pressed it onto his, giving it a firm pat, as if to say, "There you go. Thanks!" I tucked my pipes under my arm, picked up my pipe case, and headed for the door. "Come on! We're leaving! Right now!" I spoke quietly but firmly into the ears of the pipers I passed and told them to spread the word, adding, "Unless you want to walk home."

I was the driver of the minibus. They looked confused but didn't argue, although, once we got outside, they insisted on learning what was wrong. I think they understood that we escaped being discovered as frauds by the skin of our teeth. Later, of course, they said that I was the only one who'd have been beaten to a pulp, because I was a Yank. Andrew would have been sure to tell them.

18 | Caterpillars

The mated pair of moths had disjoined during the night, and in the bright morning sunshine, the male was clamped onto the net of the hanging cage to wait out the day. He was easily capable of mating again. The female was snuggled at the apex of the cage, preparing to oviposit into the folds of the fabric. Females occasionally mate a second time, given the chance, but it's not common once they have begun to lay eggs. We removed both moths from the cage and placed the female by herself into a small container, along with a bit of cotton as a site for her eggs. She wouldn't encounter cotton growing anywhere in England, but from the behavior of captive females, they do seem to prefer textured surfaces for egg laying. Cyril's group liked to use cotton, but almost anything will do just fine. I eventually adopted cheesecloth, as the eggs are much easier to separate from it than from cotton fluff.

Unlike some lepidopteran species that distribute their eggs one or a few at a time onto suitable host plants, peppered moths seem to put all their eggs (or many of them) in one basket. They might follow a crevice in tree bark for some distance, or layer the underside of a leaf with a sheet of eggs, but in captivity, the eggs are laid in just a few large batches over the course of only a few days. Also, unlike some lepidopterans, peppered moths do not feed as adults, so provisioning the eggs saps the female's tissues. Her body is quickly transformed from a robust, gravid state (swollen with eggs) to a shriveled condition. Her wings may still be resplendent, but she is unable to fly. Nonetheless, her wings may serve an important role in camouflage, hiding her from predators as she lays her eggs. Once a female is finished with this act, she loses her inhibitions about moving around during daylight hours. Healthy peppered moths never do that. It's doubtful that post-reproductive females last very long in the wild, although they can linger for a week or more in captivity.

The eggs are a pale green and remain that color for about ten days. Then they darken, turning black within a day. When I first observed this, I had feared the eggs had gone bad, because they were discolored. My worries were over the next day, when the eggs hatched into tiny caterpillars, leaving the clear, hollow chorion (embryonic membrane) "shells" behind. The cue that an egg is about to hatch is that it turns black, or, more precisely, that the larva inside develops pigment. Unfertilized eggs laid by virgin females never turn black, so after about two weeks, if the eggs are still green, one can safely conclude that they are duds.

I paid close attention to Cyril's assistants as they cared for the caterpillars. I knew this was something I'd have to do myself if I hoped to establish my own breeding stocks back home. I also understood that I'd have to modify their procedures for this to work in Williamsburg. For one thing, they used net "sleeves" in the early stages of rearing. Caterpillars feed on the leaves of living branches still attached to trees, and the sleeves kept them from wandering off. This procedure ensured a supply of fresh leaves as the larvae munched away. I wouldn't be able to use this approach for two reasons. The first—and most important—is the risk of English stock escaping into North America. Although the species is Holarctic (found throughout the northern continents of the world) in its natural distribution, thus requiring no permits to import peppered moths into the United States from the United Kingdom, there are genetic differences among geographically separated subspecies, and I, for one, didn't want British genes contaminating the local American gene pool. Another reason why I couldn't use sleeves in Williamsburg is the broadcast spraying of insecticides to control mosquitoes. I'd have to raise my caterpillars indoors, to protect them.

What should I feed them? I learned they are not fussy eaters and can be reared on a wide variety of trees. We were using birch, willow, and apple. They eat the leaves, not the wood. Not even the stems or fruits, just the leaves. It takes about a month or more for the caterpillars to fully mature before they pupate, during which time they pass through six larval instars (developmental stages). Each successive instar is larger than the preceding one, and to allow this growth spurt, a larva must shed its "skin," which is made of chitin and doesn't expand. It's like a suit of armor on a knight and is, in, fact the insect's external skeleton (exoskeleton). (This is true for all members of the animal phylum, or taxonomic classification, Arthropoda.) The instars are separated by molts, meaning the exoskeleton is shed, the larva's size expands, and a new exoskeleton is secreted to replace the old one.

Growth happens before the new exoskeleton hardens. There is a good bit of stretching between the body's plates, but growth is otherwise restricted within an instar.

This is all fairly basic biology to people familiar with insects. But what caught my eye as the larvae grew was their astounding resemblance to the twigs of their host plants. One had to look hard to find the larvae, as they remained dead still during the daylight hours and precisely matched the color of the twigs. If they all managed to look like, say, birch twigs, that might be amazing enough. But the only ones that did look like birch twigs were the ones fed on birch leaves. Those fed on willow looked like willow twigs. And, those fed on apple looked like apple twigs. How did they do that? Was it something in the leaves they ate that triggered the development of their colors, or was the stimulus visual—that is, did they match the colors they saw? These were questions I would invite several of my future research students to explore.

During the final instar stage, larvae are about as thick as a pencil and a couple of inches long, and they eat huge quantities of leaves daily. Ultimately they drop to the ground and burrow into the leaf litter or soil, where they pupate. In the lab we provide them with dampened peat moss to serve as a pupation medium. The puparium is shiny dark brown, rather scat-like in appearance, and is not surrounded by a cocoon of silk. Peppered moths spend the winter in the pupal stage, underground. In the spring, adults emerge (eclose) from the puparia, completing their life cycle. My plan was to take a bunch of pupae home with me, as I intended to keep the moth stocks going. The genetic cross we made would produce both adult phenotypes (fully melanic *carbonaria* and pale typicals), and I had a ready source of suitable host plants in Williamsburg.

In my earlier research with fruit flies and my more recent stint with parasitoid wasps, I would get new generations every two weeks or so, depending on the species and the rearing conditions, but with peppered moths I would have only one generation per year. Small wonder that serious geneticists choose to work with *Drosophila* rather than *Biston*. They have other good reasons, in addition to life in the fast lane.

19 | Long Season's End

"That's what wives are for," Cyril laughed, as we discussed how best to "smuggle" the pupae from the United Kingdom into the United States. Although transferring *Biston betularia* between these particular countries is unrestricted, as the species' natural distribution includes the entire Northern Hemisphere, those entrusted with guarding international borders don't necessarily know that, and they most likely would not know how to identify a species by looking at pupae. So, even with official US Department of Agriculture (USDA) paperwork in hand, stating that peppered moths may be brought into the country (providing no host plants or soil accompanied the specimens), delays at customs entry points were routine. Worse, living materials were usually confiscated for further examination. Once the identity stated on the USDA documentation was confirmed by their experts, the items were then returned, but by that time everything would be dead. Dead! Well, we weren't running a museum. We needed live stuff. Transferring living insects across borders was a tricky business, even when it was perfectly legal. We didn't want to break the law, but presenting specimens to border guards for inspection seldom ended well.

Cyril went on chortling, as only an otherwise charming old man could get away with. "That way if your wife is arrested for smuggling, you're still free to carry on with the work. You mustn't put yourself at risk!" He thoroughly enjoyed this sort of humor, affecting a cavalier disregard for conventional politesse, but in this instance, I wasn't sure he was joking. Lady Clarke smiled knowingly, having no doubt smuggled her fair share of bugs into England. Cyril imported exotic species from the far-flung corners of the world, but most of this traffic was through the mail.

My wife Cathy had arrived with our daughters in mid-July. Her plan was to spend a fortnight with us before returning to her job in Williamsburg. The girls would stay with me in England until we headed home together, in time for the start of school at the end of August. The last peppered moth we caught in 1984 was July 28, effectively ending the flying season and my summer's work. Caterpillars still had to be fed, but that task fell to Cyril's assistants. I was merely an observer. I was also the gracious recipient of a small, pocket-sized box full of new pupae that I gave to Cathy to carry home for me, in exchange for my watching the kids. That seemed fair. Cyril and Lady Clarke liked Cathy straight away for being such a good sport about this. I'd known her for far longer and wasn't at all surprised. She'd always been a tolerant sort and a risk taker.

We went to London by train a few days before her flight back, to take in the sights that rank high on tourists' lists, such as the Tower of London, Buckingham Palace, and the British Museum. There are enough travelogues about what a super place London is to visit that I needn't repeat that information here. I do, however, have a few personal yarns.

On one occasion, we four were riding on the top deck of a double-decker bus. (We always rode on the top deck.) We were planning to get off at Charing Cross, but we weren't sure we'd recognize the stop when we came to it, so we asked the conductor to please tell us when we arrived at that locale. From time to time the conductor would come up the stairs when making his rounds, and at every single instance we'd look toward him in anticipation as his head appeared in the stairwell. He'd look back and say in a thick Pakistani accent, "No! No! Not yet Charing Cross!" Then he'd go on about his business. This happened at least a half dozen times. When the bus finally did reach Charing Cross, before the conductor could make it up the stairs to announce it, the entire crowd of passengers riding with us on the upper deck shouted in virtual unison, "CHARING CROSS!!!"

On another bus ride, we were looking for Saint Paul's Cathedral. Cathy saw it from the bus window as we passed by, so we signaled our intention to get off at the next stop. Elspeth, our nine-year-old, also spotted a McDonald's from the bus window, and it was lunchtime. Where to go first? During my stays in England, my favorite fast food is fish and chips. Then and now. I could (and too often did) eat fish and chips every day. But as none of us had been inside a McDonald's outside the United States, we thought it'd be fun to compare a British McDonald's with our familiar American ones. By the time we got off the bus, we had become disoriented about which direction led to Saint Paul's and which to McDonald's. Cathy spotted a bobby, complete with the distinctive tall helmet, and decided to ask him. She knew better than request me to ask someone for directions, so she queried, "Could you please tell us where Saint Paul's Cathedral is, or where we might find a McDonald's?"

He paused for a long moment and actually scratched his head before answering: "Well, from here Saint Paul's is a bit trickier to find than McDonald's."

The following day we arrived at Gatwick, to merge into the nastiest congestion we'd ever encountered at any airport. Cathy's return flight on Virgin Atlantic had been heavily overbooked. There must have been 900 people in the queue for the same flight. At the risk of sounding like Dave Barry, I'm not making this up. I loathe hearing wearisome stories about airlines, overbooked flights, being stuck in airports, and on and on, so I'll spare you this one. Any flight you survive, arriving alive and in one piece, is a good flight. Enough said.

Elspeth sobbed openly when her mom's ground transportation left Gatwick for Heathrow Airport, to catch another plane. Megan and I felt the same way, but we were too big to cry. Then we three went back to Euston Station to catch a train for Lime Street Station, Liverpool.

20 | Yankees Go Home

We three seized our free time in August to become tourists. On Scottish soil, I piped for my daughters to dance the fling. No busking involved. Just for the pure joy of it. Megan later competed in Glasgow at the Worlds, and over the years she's been back many times as a dancer, teacher, coach, and judge. But this was their first real visit to that part of the world, and we took it in my old yellow-and-black Hillman Hunter, watching the road whiz by under the floorboards.

We also went to visit cousins in Cumbria, the most stunningly beautiful part of England. My father was born there, in the coastal town of Whitehaven, but he was whisked away to America by his widowed mother following the Great War. I was always amazed that he could remember so much about the place, as he was only nine years old when he emigrated. We delighted in combing the same hills and seacoast and hearing tales from a different time.

But mostly we stayed in our Birkenhead flat, taking day trips and interacting with the City of Liverpool Pipe Band. Joe and Toni Aspinall and their three young daughters lived just a few blocks from us. They were delightfully sociable and warm people, and they practically adopted us into their family. My older daughter Megan is very outgoing and friendly, so they got along especially well. At 17, she was around the same age as several of the band members, so she got to go with them to the haunts of Liverpool where the Beatles cut their musical teeth. Elspeth, being much younger, didn't get to spend time in the pubs, but Pip, the teenage drummer, took pity on her and gave her a pair of outlandishly fluorescent bobby sox to acquire the Mersey look.

It was actually Pip who inspired my response to the Glasgow inquisition about my nationality (see Chapter 17). Earlier that season, on a band outing, Pip proudly informed me that she was portraying an American in a school play, and to do that, she had to mimic an American accent. When I asked her to say a few words for me, she responded, in her usual Scouse accent, with "Can't be bothered." I continued to pester her for a sample, telling her I might be able to offer suggestions if she'd like some coaching. Pip continued to say "Can't be bothered," but she finally gave in, intoning "The rain in Spain stays mainly in the plain." I was flabbergasted by how beautifully she spoke these words. It didn't at all sound like an American accent, but resembled something straight out of *My Fair Lady*, in an English accent so "proper" that Julie Andrews would be envious.

I stated as much to Pip. She apologized, saying she couldn't conjure up her American accent but instead slipped into her "telephone voice." Telephone voice? She explained that her mother insisted she speak like a BBC newsreader when answering the telephone. I thought I understood and asked if her telephone voice was the one she used at school, while her Scouse speech was relegated to the streets. "Oh no," she said. "I could never use me telephone voice at school. Me mates wouldn't allow it. They'd think I'd gone posh."

"Posh," incidentally, is an apocryphal acronym for "port out–starboard home," which was printed on premium tickets for return (roundtrip) sea voyages. It meant that the best cabins were on the lee rather than the windward side of the ship. These were by far the most expensive accommodations, so "posh" was reserved for the elites. For members of the working class, an accusation that one has "gone posh" is the ultimate put-down. It means that person is a phony, trying to deny his or her own heritage by assuming the accent and manner of speech of the upper classes, so as to move in their circles. Putting on airs. Upward mobility is not admired by those left behind. While I'd assumed that people were kept down by those above them in the hierarchy, I learned from Pip that the strongest disincentive to rising above your station is peer pressure from your own mates.

There was no question that the members of our bagpipe band would have felt unwelcome in Sir Cyril's sailing club, unless they were hired as day workers to scrape barnacles. Nor would Sir Cyril have been accepted as a member of that band. Their cultures were as immiscible as oil and water. They were separate and unequal. While class distinction in the United Kingdom wasn't what it had been generations before, it was still very real. Yet I and my family moved freely through it, at least at certain levels and to a limited degree, probably because we were Americans and too unsophisticated to know our places. We were humored, if not accepted.

I learned a few things about English subculture that summer that my father didn't know. My time spent among these quirky people was nearly always pleasant, and it ended too soon.

I briefly discussed what to do about my car with Tommy Graham. It wasn't worth much, but rather than sell it, I wanted to give it to his son Andrew. Andrew rode a motorbike, rain or shine, and with bad weather not far in the future, I thought he might prefer to get around inside a car. Tommy wouldn't hear of it. He looked this gift horse squarely in the mouth. "I won't have you put my son in that death trap! The M.O.T. is about up, and it'll never pass. You can see the road right through the floor!" And on and on he went about the problems with my vehicle. Andrew had no say in the matter. Tommy forbid me to even mention it to him.

So I decided to give the car to Joe. He had already accepted a job in South Africa, but as he and his family would not be emigrating until November, Joe figured he could squeeze a few months' use out of my old banger. That would save him having to borrow one whenever he needed wheels. The day before my daughters and I left our Birkenhead flat for the final time, I signed the vehicle over to Joe. He agreed to drive us in his "new" car to Lime Street Station to catch the morning train to London.

The night before leaving, I decided to throw myself a farewell party. I invited all of the band members to come to my flat for beer and munchies. We got a bagpipe tuned up and passed it around, making a happy din. At some point Tommy presented me with a handmade wall plaque, shaped like a shield on which was painted a picture of a stylized Liver Bird, the symbol of Liverpool and, appropriately enough, the image adopted as the symbol of the pipe band bearing its name. Below the Liver Bird was a small copper plate inscribed with my name and band rank: piper. I was puzzled by the additional words "Ne Oublie." Tommy explained this was the motto of his own Clan Graham, meaning "Never Forget." He promised the band would never forget me, and he hoped I would never forget them. He quickly added a touch of Scouse humor to this otherwise sentimental moment: "'Ne Oublie' is an especially appropriate motto for the band, because we must first memorize the music before we put it on the pipes. *Never forget*, or you'll go off!"

Now it was my turn to hand out gifts, so I invited them to cart away whatever they wanted among the household items I had purchased during my three months living there. Anything not nailed down was up for grabs.

For example, I had bought a black-and-white TV from a pawnshop for 10 quid. I didn't watch it much, but it helped me keep up with the news and soak up some of the culture. I only learned as I was giving the TV away that I was supposed to have a license for it.

"A license?" I asked incredulously. "What for?" I was informed that this was how they funded the BBC. The consumers of the product paid for it. I had never heard of such a thing until then, although now I think it's a great idea. It's certainly a big improvement over the begging American PBS stations do during their seemingly endless fundraising campaigns. Even when you make a pledge, you still have to endure the near-constant pleas until they reach their goal. The British system seems much better, but at the time I didn't think so. I was skeptical when they told me that officials patrol the streets in special vehicles equipped with electronic surveillance gear to catch people who are watching TV without a license.

"Oh, come on!" I said. "How can they tell if you are *receiving* a signal? Certainly they can tell if someone is *transmitting* a signal, but how do they detect the reception of a signal unless the TVs in this country have beacons that announce when a TV is turned on. Hmm? Is that how they do it? Well, if they don't want me to intercept their broadcasts, they should keep their signals out of my living space. Isn't this the country that claims a man's home is his castle?"

Tommy, by now fully recovered from his lapse into sentimentality, countered my harangue the way he ended all arguments with Americans since World War II: "That's the trouble with Yanks. They're overpaid, oversexed, and over here!"

The other items I gave away didn't generate that much discussion. Most of the stuff was from the kitchen: pots and pans, food, and the like. I even gave Big Kenny my Andy Capp–style cap. I thought it made me look British, but Tommy said it had "Yank" written all over it. Kenny was happy to have a hat that made him look like a Yank. At the party's end, the band members poured into the street carrying away their booty. It looked like a burglary in progress. No worries there. Some bloke lugging a TV down the street in the middle of the night wouldn't arouse curiosity in Birkenhead.

The next morning, right on schedule, Joe was there with the car to take us to Lime Street Station. We were all packed up and ready to go. I was relieved that Joe was driving. I never enjoyed wending my way through the busy tunnels from Birkenhead into Liverpool, and this was still morning rush hour. Joe was fully relaxed and full of gab.

As we walked toward the platform for our waiting train, the origination point for travel to London's Euston Station, I thought I caught a glimpse of Big Kenny out of the corner of my eye. How could that be? What would he be doing here? But yes, it indeed was Big Kenny. At first he pretended not to see me and attempted to conceal himself among the normal-sized people scurrying to their trains. Then he smiled awkwardly and acknowledged being noticed in the crowd, as if to say, "Fancy seeing you here." Hmm? Something's up, I mused. Surely Big Kenny wouldn't come down here all by himself to wave goodbye. We just saw him a few hours ago. And he's still wearing my hat! We weren't near enough to each other along the platform to speak, nor was there time to work my way over to him. He just looked a bit sheepish as we waved and got into the coach with Joe ushering us along nervously.

Then we heard it. The sound was unmistakable. Everything became clear in that instant. It was the sound of bagpipes coming down the platform. The City of Liverpool Pipe Band had come to pipe us off at the train station.

Well, it wasn't exactly the entire band. Just four members were there, not counting me. But it was enough to produce the most vivid memory and biggest honor I have ever received, Scouse honor!

The people seated around us in the coach must have thought Prince Charles was on the train. Who was possibly so important that a pipe band would show up at Lime Street Station? When the band did march up to our car, its modest size suggested that the pipers weren't there to play for the heir to the British throne. The band members at the station included Tommy and his son Gordon, both dressed to the nines, with full plaids, horsehair sporrans, spats, feather bonnets, the works! They were totally resplendent. They had to have been up for hours to put all that gear on and then get tuned up to pipe us off at the station.

Well, we couldn't just sit there in our seats while they piped their hearts out on the platform just outside our window. So we got up and went to stand next to them on the platform as they piped. There were still a few minutes before the train was to depart, and the conductor nodded to us that it was OK, that he'd warn us when we'd have to get back on board. By now a small crowd had gathered around, and we were just part of that group. When

Tommy and Gordon finished their set, I decided to engage in a little Scouse humor of my own. I called out loudly to Tommy, "Are they hard to play?"

"No," he responded, picking up instantly on my lead for the gag we were about to pull on the unsuspecting group of onlookers. He lifted the pipes from his shoulder and extended them to me. "Try 'em. Give us a blow."

So, I took the pipes from him and immediately began playing "Gathering of the Grahams" (to honor Tommy, of course), a 6/8 march more familiarly known as "Atholl Highlanders." Before I got through the first part, Gordon joined me in a tight duet. When we finished the tune, I took the pipes from my shoulder and handed them back to Tommy. "You're right, mate. They are easy to play. Ta."

By this time the train conductor motioned to us to get back on board, so with broad smiles and hearty waves, our train pulled out for London with the fading sound of bagpipes biding us farewell. No doubt our fellow passengers were wondering "Who in hell are these people?" but they were too British to ask. We had become too British to tell them.

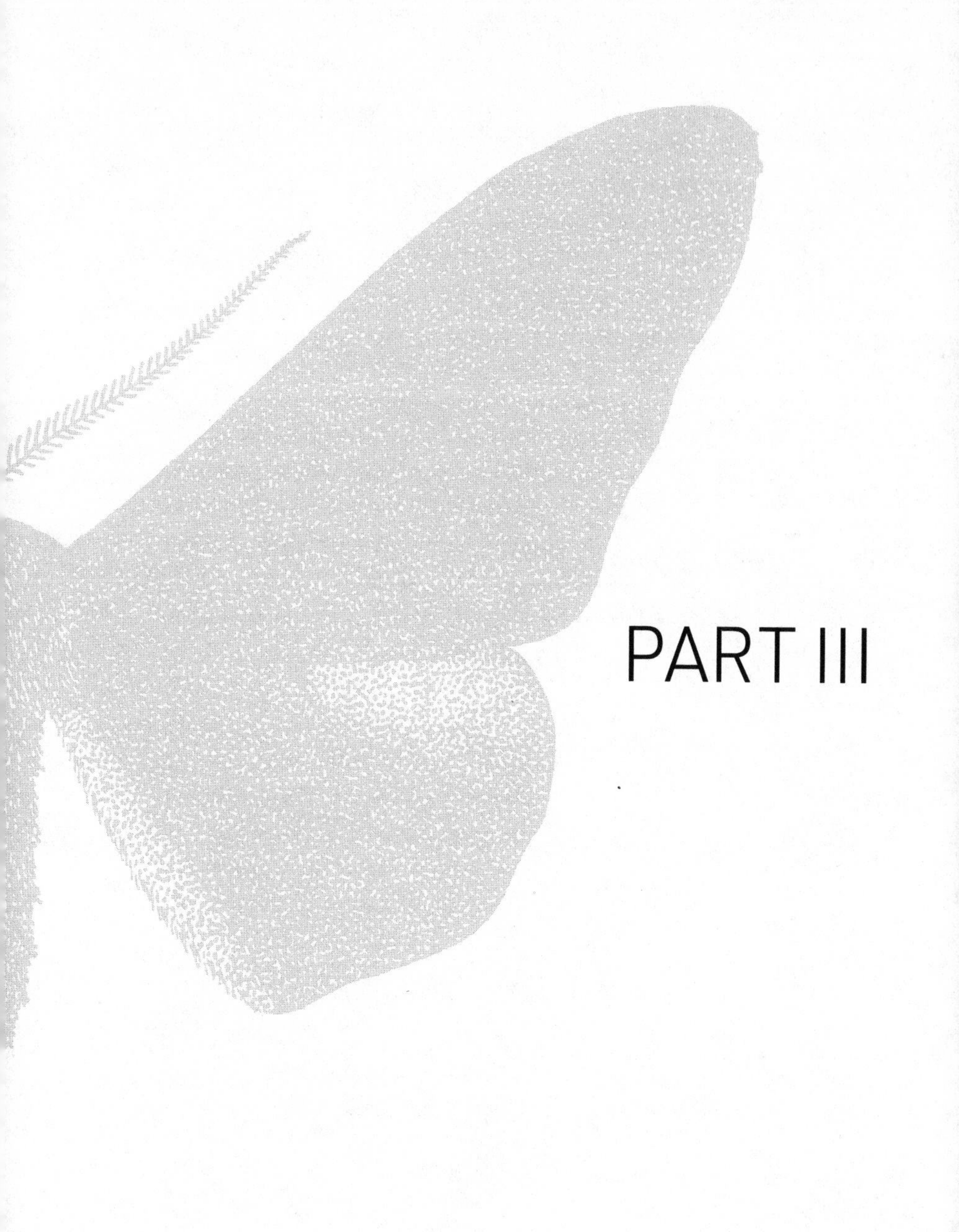

PART III

21 | From Field to Lab

Back in the United States, in southeastern Virginia, at the venerable College of William & Mary, on a table in my lab on the third floor of Millington Hall, sat a small box that could easily be tucked into a shirt pocket. It was the box containing the peppered moth pupae my wife Cathy carried back from England. David West, standing behind the box, advised, "I should put them in the fridge."

Dave was visiting from Blacksburg, at my invitation, to present a departmental seminar on his latest work with butterflies and, not incidentally, to pick up his share of the moth pupae. He would hybridize a sample of *carbonaria* adults emerging from this brood with the American subspecies *Biston betularia cognataria*, making reciprocal crosses in an attempt to test for dominance of the *carbonaria* allele on a naïve genetic background. (This topic will be discussed in Chapter 37.) The remainder of the imported pupae would stay with me in Williamsburg, to set up a breeding stock of British peppered moths.

As this brood was produced by a *typica* female mated with a *carbonaria* male, half of the progeny were expected to be fully melanic *carbonaria*, and the other half pale typicals, assuming the male parent was heterozygous for melanism (having one *carbonaria* and one *typica* allele); or, if the father had been a homozygote for melanism (having two *carbonaria* alleles), *all* of the progeny would be fully melanic, heterozygous *carbonaria*. Either way, I would use them in rest-site selection experiments, as well as make crosses to get additional generations, so I could continue trials in my lab under controlled conditions.

My more immediate concern was to prevent the premature emergence of adult moths from these pupae in late fall or the dead of winter, when there would not be fresh host plant material available to feed the offspring. We needed to delay eclosion until spring. Dave suggested storing them in the refrigerator, simulating winter conditions, until the trees outdoors put out a good supply of fresh leaves.

Cyril Clarke stored his pupae outdoors in peat moss, under more-or-less natural conditions, so they would remain in developmental synchrony with the local wild population by experiencing identical seasonal cues. But warm Williamsburg seasons are not like those of cool northern England, including having pronounced latitudinal differences in day length. While I recognized that refrigeration would postpone emergence until the pupae were warmed up, I worried about exactly when I should plunge them into the cold. At what point would they have developed sufficiently within the puparium to be ready for artificial winter conditions? Were these pupae already in diapause (a hibernation-like pause in their development)? Or *do* they diapause? If so, have they received the appropriate environmental cues to stimulate this state? If I placed them into the fridge too early, would that kill them? If I waited too long, would they emerge before I could put them there? Or, what if they were about to emerge just as I was sticking them into the cold? I was a nervous wreck about what to do. I didn't want to lose these pupae. My whole research project depended on their surviving my treatment of them. I was without experience in this phase of their husbandry, so I hoped Dave would offer some guidance. To his suggestion that I put them in the fridge, I asked: "*When* should I do that?"

He paused for a long moment, raised his eyebrows up and down a few times as he seemed to be pondering elapsed time since pupation, then said, "Now."

My next bit of concern was to maintain them in a healthy condition. I kept them in dampened peat moss, just like Cyril Clarke had shown me. When I first asked him exactly how moist that damp condition should be, he held up the box of pupae and told me to put my fingers into the peat moss. It was easier for him to demonstrate the feel of the stuff than to explain it. My fingers remembered, and still do. I can tell what is too damp and what is too dry. Both deviations from just right are lethal for the pupae. Peat moss is funny stuff in that regard. Too dry, the way it comes straight from the sack, and it'll draw the life juices out of a pupa, killing it. Too wet is just as bad, because the pupa will still die, perhaps by drowning or suffocation, and this is rapidly followed by mold growing on it. It's easier to tell a dead pupa with mold on it than one that has simply dried out but otherwise looks healthy. One way to determine whether pupae are alive or dead is to hold them in the palm of your hand and close your fingers ever so gently around them, so as not to cause any injury. They react to this by squirming, and if you are holding several of them, it is very obvious that they are alive. It is not a good sign at all when a bunch of pupae held in the hand remain motionless. I also learned that little trick from Cyril, and it gave him great pleasure to pass on the secrets of *Biston* husbandry.

So, to keep the pupae healthy in the fridge during the 4°C "winter," I would visit them daily, stirring up the peat moss gently with my fingers, so it would not develop lethal wet spots. This would serve to aerate the bedding and would give a hint about whether to mist-spray a bit more water onto the peat. I was nervous about all of this, having had no previous personal experience. Over time I developed McClintock's feel for the organism, and that first year I fortunately hit it just right, thanks to the coaching of Cyril Clarke and David West. You can't learn this stuff from books—except maybe from this one.

During the next several months, while I waited for the arrival of spring and the 1985 moth season, I constructed an improved test pen, designed to reduce overlaps in scoring moth resting positions, and I replaced the white stripes with a moth-tested shade of gray. The new pen had alternating half-meter-wide, black and gray stripes on each of four square-meter walls. The contrasting surface reflectance of the uniformly textured fabric walls of the pen measured 2.5% and 50.4%, compared with a barium sulfate standard white. Initially, I planned to test moth rest-site selections under conditions of natural lighting. I set up a screened dining canopy on the roof of the biology building, to protect the test pen from rain and serve as an additional barrier to the moths' escape, but every strong storm that came along threatened to blow the whole contraption off the rooftop. I also worried that insecticide sprayed outdoors to control mosquitoes might also kill my precious moths. I ultimately decided to move the entire operation indoors, but not before comparing the behavior of moths kept in test pens under both indoor and natural lighting.

By this time I had constructed similar moth pens to test males and females separately. To reduce attraction between the sexes, these pens were isolated in two closed, windowless, air-conditioned laboratory prep rooms, separated by four stories: one in the basement, and the other on the third floor.

I installed incandescent lamps, controlled by dimmer rheostats, above the pens, so I could simulate dawn and dusk. To approximate how gradually to raise and lower the lights to create artificial "night" and "day," I used a stopwatch and a light meter to time real sunrises and sunsets. I was painstakingly conscientious about this and never just flipped the lights on or off to announce dawn or dusk. As far as I could tell from my trials, the moths reacted to the simulated and natural sunsets and sunrises in the same way.

This setup allowed me to run short-day experiments, in the hope I might squeeze more choices out of each healthy moth as it emerged into 12-hour days with alternating 6-hour light/dark cycles. The experimental design was to "ask" each moth the same question over and over again, to see if it gave consistent answers about its background preference. Individually marked moths were put on the floor of a test pen each night, and their resting locations (on the black or gray stripes) were recorded each morning. I wanted to learn whether their "choices" regarding black versus gray were random. Or would all of the moths show a statistical bias for the same surface reflectance? Instead, might some of them choose gray repeatedly, while others chose black? I could think of no better way to find out than by leaving it up to the moths.

To avoid overcrowding the test pens with active moths, I spaced out their emergences by removing just a few pupae from the fridge every several days, beginning on Saint Patrick's Day, March 17, 1985. In about 10 days to two weeks, the pupae removed from the cold begin to "hatch." The sexes from the same brood are on slightly different developmental schedules, with the females emerging first, often by several days. There are exceptions, and the timing for some males and females overlaps, but in nature, females very probably disperse well away from the natal site before their brothers emerge. Different maturation schedules between the sexes within sibships are the norm among the disparate insects I have worked with (*Drosophila* species and parasitoid wasps). Whatever the developmental reasons for this, it would reduce the likelihood of close inbreeding in the wild. Once started, emergences from peppered moth broods peak within a few days and finish after about a week or so. After that, any unhatched pupae are probably duds.

By April Fool's Day I was in business, and the several batches of pupae from this single brood kept me testing adult rest-site selection until the middle of July. But the work had not ended then. To carry on with the project, I needed to breed these captive moths and raise their progeny, to learn if background choice preferences might be heritable (have a genetic component). This meant providing caterpillars with fresh leaves, which they consumed in great quantities once they achieved an appreciable size. This daily chore of feeding caterpillars continued through September, with some stragglers still demanding food into October! As the caterpillars continued to grow larger, they required much more space. I used 40-gallon trash barrels as "vials," into which I replenished their supply of fresh leaves, still on branches. The bigger they got, the faster they stripped the leaves from the branches. Caterpillars in their sixth (and final) instar stage are a trim couple of inches long (60–70 mm), yet, even with hundreds of them in a barrel, right in front of your eyes, they are hard to see when they remain motionless among the twigs and branches of their host plant. I had these feeding barrels in my tiny research lab, which I shared with my three graduate students. They were all heavily engaged in master's thesis research on parasitoid wasps (*Nasonia vitripennis*), and they were feeling a bit cramped after I jammed my caterpillar barrels into their room. One of them, the then young Allen Orr, made this bold prediction: "Well, boys, once we're gone this lab is going totally *Biston*!" Allen enjoyed looking at the striking resemblance the *Biston* caterpillars bore to the twigs in the barrel. Impressed by their nearly perfect camouflage, he boomed: "Don't tell me natural selection doesn't work!"

He even brought in people from the hallways to show off our livestock. One was my forthright, outspoken colleague Charlotte Mangum. She thought we were pulling her leg by claiming there were caterpillars in the barrel. "All I see are twigs," she insisted. So I poked one of the caterpillars to get it to move. Suddenly scores of them became visible as her eye adjusted to seeing them. "Ooh! Now that is amazing," she said. Everybody had similar reactions to the caterpillars' astonishing crypsis.

Allen's prediction about the direction my lab would take came true. As I phased out wasp research over the next several years, peppered moths dominated my operation, including the remarkable mimicry of the caterpillars to the twigs of whatever food plant we provided for them. That peppered moth caterpillars developed different colors was known at least a hundred years before the first melanic adult moth was reported in the literature. But the two phenomena are unrelated.

Young caterpillars can end up on a variety of trees in mixed deciduous forests, depending on which way the wind blows them when they first hatch and balloon away, suspended on silk threads. Once settled, they develop colors that match the particular tree twigs they happen to have landed on. It's not a quick change, like a chameleon or a cuttlefish might accomplish (with special cells called chromatophores) to quickly blend into backgrounds they frequently move across. Instead, the colors of peppered moth caterpillars involve the synthesis of pigments deposited in the body wall, a much more gradual process. My graduate student, Robin Parnell, showed by extensive controlled experiments that it is what these caterpillars saw, rather than what they ate, that provided the stimulus. My undergraduate honors research student, Mohamed Noor, painstakingly demonstrated that this visual response is reversible until the final instar—that is, a green caterpillar can become brown, and vice versa, when transferred to different backgrounds during development. More recent work by Ilik Saccheri's group in Liverpool has shown that their visual stimulation is not restricted to larval eyes (ocelli). There appears to be extraocular photoreception, although the mechanism has not yet been identified. Remarkably, caterpillars are also attracted to the twigs they match via extraocular photoreception.

Now back to the behavior of the adult moths. The short-day experiments meant long days for me. In addition to simulating dawns and dusks twice a day in two separate rooms, plus scoring the moths' background selections, I managed to make close observations of the moths' nighttime behavior. I could see into the pens in the dim, simulated starlight, but just barely. I was also able to watch other moths that were put into mating cages in my office. In addition, I kept several moths in cages that housed artificial trees made from wooden dowels. As Yogi Berra put it: "You can observe a lot by watching."

For one thing, moths don't waste energy. Lazy? No. Efficient? Yes. Let's start with the females. The first night following eclosion, they are very active. In the test pen, they begin flying shortly after "nightfall" and keep up a constant motion, banging off the soft walls of the pen for several hours.

They orient toward the source of light, however dim it might be, in a darkened (simulated night) room. They seem in a frenzy to escape the confinement of the pen. By the wee hours of the morning, they have calmed down considerably and stand erect, with their wings held high above their backs, antennae elevated, trembling just perceptibly. It is not at all clear from this posture whether a moth can compare the contrast of its wings with the surface reflectance below. If so, using either paint or paper collars to conceal the circumocular scales, which Kettlewell proposed as being key to background matching, would be useless. In this nearly motionless posture, the females seem to be releasing pheromone from a genital apparatus protruding conspicuously from an extended abdomen. At dawn they generally clamp down wherever they are, having selected their daytime resting position at night, in the dim starlight, rather than waiting for the sun to light the way.

Kettlewell described this clamping behavior in considerable detail. I would add to it and emend his interpretation of its function. A peppered moth at rest on a branch (or a wooden dowel substitute) holds its body in line with the orientation of the branch during the night, with its wings held high above its back, but not as high as butterflies do when they alight. This is the "calling" posture—at attention and fully alert. At dawn, the moth begins to pump its wings up and down as it turns from side to side. At first the wings pass by the branch without making contact, so up they go again. But as the moth continues to rotate its body, the wing tips make contact with the surface of the branch. Finally, when its body is oriented at a right angle across the branch, the wing tips make full contact with the branch surface and the moth comes to rest, clamped in that position. Rather than exploring the surface in a reflectance-matching search, as Kettlewell proposed, it seems far more likely that this behavior is involved with body orientation along a branch. This is consistent with Kauri Mikkola's interpretation, based on his observations of captive specimens: peppered moths rest along the undersides of branches. Yet even on comparatively flat surfaces, such as the trunk of a tree or the wall of a large cage, the moths don't seem to explore very far from their nighttime calling positions as dawn breaks. They do turn a little this way and that, as if gauging the surface contour, and pump their wings slowly and deliberately, similar to doing pushups, but they take very little time to settle in for the day. The whole exercise lasts about 5 to 10 seconds. Clamping down has more to do with geometry than with reflectance.

My observations of captive females show them to be extremely active their first night of adulthood, but on successive nights, they very quickly settle into calling postures. What are they calling? Males, of course. They do this by releasing pheromone, an attractive scent males can't resist. Perfume! The males home in on it to find a receptive virgin female. After her initial dispersal flight away from where she emerged, a female tends to stay put. It won't do her much good to call out "Here I am, boys" and then take flight to some other location. What would be the sense in doing that? So, night after night, after that first night of hitting the road, a female will call out for a mate. Once she has mated, she never flies again. She simply lays eggs and then dies.

I did discover that if a virgin female is left where she is found in a test pen, she'll stay in that exact spot night after night and not readjust her position. Well, this isn't as precise as physics. Maybe she moves a little bit, but basically she stays put. If, however, I put her back onto the floor of the test pen before nightfall, she'll move back up onto the walls of the pen under the cover of darkness and take up a new resting place. Would it be on a surface of the same or a different reflectance? Do females show preferences for background reflectance, or are their choices random?

I'll hold you in suspense a wee bit longer, in order to tell you about male behavior. What do they do in the dark? During the daytime, as we know, they do just what females do: they remain dead still against whatever background they happen to be on, in an "I'm not here" pose. Their wings are pressed flat against the surface, and if they match that surface, they really are hard to see. Only motion will betray them to most eyes, and unless they are discovered and vigorously disturbed, they refuse to move. At night, it's different story. They are active and move about, but, unlike females, they do this for the rest of their adult lives, not just the first night. In captivity, they can easily be kept alive as adults for a week or more. But eventually they run out of gas. They don't feed as adults, so their bodies start to shrivel. They also show other signs: wear-tattered wings and a loss of scales being the most obvious. For the first few days, however, they go full throttle. I wanted answers about rest background preferences from these guys. Do they show any?

When stimulated, male moths bang around the test pen all night, and they will do it every night until they die or become decrepit. But what stimulates them? Pheromone, obviously! Deprived of that mighty attraction, they are couch potatoes. They are certainly on full alert at night and will quickly zoom away if annoyed. I had been under the impression that male moths routinely fly around at night in active searches for females—cruising for chicks like teenage boys in a convertible. I recall a seminar by an eminent biomechanics researcher who projected scanning electron micrographs of plumose male moth antennae to target his abstruse calculations of wind-shear forces across the antenna combs. He depicted male moths as flying around with their antennae held out to catch molecules of pheromone and then homing in on the source. Well, maybe. More likely, what happens is that males come to their "attention posture" at night, lifting their wings like weather vanes and holding their antennae high, the latter spread to catch incoming pheromone gently wafting on the breeze. Only after they've caught the exact scent do they take flight, following the gradient toward its source. Although wind-shear forces would kick in either way, males don't just fly about willy-nilly, looking for females. They sit patiently and wait to be called. Saves gas!

Otherwise, without any motivation to move, males just stand in place pretty much all night long and then clamp down at dawn to ride out the day. At night, back up go the wings and antennae. They'll do this night after night, without changing their position very much at all, except for reorienting their body axis along a branch. It would not do to score the same moth each morning in a background selection experiment, exclaiming, "Oh look, he's on black again!" That would be like flipping a coin just once but coming back to examine the same coin each morning, crowing, "Oh look, another head!" You could easily get 10 heads in a row that way, but that would be cheating.

Clearly, the only way to do background selection experiments correctly when using the same moths over successive nights is to force them to make a new decision each night. To accomplish this, the moths were always returned to their starting position on the floor in the center of the test pen before each run of the experiment. The next morning, their rest-site positions (either on black or gray) were recorded. Moths remaining on the floor were not scored as having made a choice (usually these moths were in bad condition and were removed from further testing). Those moths overlapping contrasting backgrounds were not recorded as having made a choice. Each moth's history throughout its time in the test pen was identified by a tiny numbered paper tag, glued to a wing by a dab of nail polish. Moths didn't seem to be aware of the tag once the adhesive dried, but they would attempt to shake it off initially. The tags were put on early in the morning, just after eclosion, and each moth's life from then on was followed until it died.

So what did I learn from all this? What did the moths tell me? Obviously, a test pen, no matter how it's constructed, is not the real world. It is a cage! Like most caged animals, the moths spend most of their time trying to escape from it. If I were to have removed the lid of the pen at night, all of the moths would have flown away, except for the ones languishing on the floor. Instead, I kept them inside and forced them to take up resting positions at dawn. This they did. They clamped down each morning. So, given that the surface area inside the cage was divided equally into light and dark bands, were the moths indifferent to those reflectance backgrounds? The answer is most definitely not. Statistically, if the moths settled randomly on the backgrounds, such results would be clear. The moths settled nonrandomly—at least some of them did.

Indeed, some moths settled on black far more often than on gray, while others did the reverse, much more frequently choosing gray than black. Moreover, additional moths didn't seem to show any bias for one or the other shade. We might offer the very same qualitative descriptive for random selections, and we could predict from binomial calculations just how often certain results are expected. To illustrate how this works, let's consider flipping coins. Let us assume both the coin and the technique of flipping it are fair, so the probability of a head or a tail appearing is exactly equal on each flip: 50:50, or 50% each, or 0.5 to 0.5, or 1:1, or ½:½. Exactly equal! We can then apply the binomial $(a + b)^n$ to describe the entire frequency distribution of possibilities of heads to tails in n flips of a coin. Say we decided to flip a coin 10 times and call that a single trial run. What is the likelihood of getting 10 heads in a row? That's an easy one. We don't even need to use binomial calculations for that (although we are using one extreme end of it). Well, the probability of getting the first head is ½. Since our assumption is that both the coin and flipping technique are fair, and that the coin has no memory of what had happened previously, we can assume that the probability of a second head appearing on the second flip is also ½. As these two flips are independent of each other regarding their separate outcomes, we can apply the law of independent events, which says that the probability of independent events occurring together is equal to the product of their separate probabilities. Ergo, the probability of 2 heads in a row (or 2 heads at the same time, if 2 separate fair coins are flipped) is equal to $½ \times ½ = ¼$, or 0.25, or 25%. So we can expect 2 heads in a row 25% of the time, while some other result (not 2 heads in a row) will occur 75% of the time. By the same logic, we can extend this to ask what is the probability of 10 heads in a row? It's $(½)^{10} = ½ \times ½ \times ½ \times ½ \times ½ \times ½ \times ½ \times ½ \times ½ \times ½ = 1/1024$, or less than 1 in 1,000 times you flip a coin 10 times in a row. Meaning 1,023 times out of 1,024 tries (99.9% of the time), you'll get something other than 10 heads in a row.

OK, now for something completely different. What are the chances of getting 10 tails in a row? Well, the logic and algebra are identical. Again, it would be 1/1024. We could jazz it up a little by asking what are the chances, if you flipped a coin 10 times, that you'd get *either* 10 heads in a row *or* 10 tails in a row? Basically, you are asking what is the chance of getting 10 outcomes that are the same, when it doesn't matter if all the coin faces show as heads or tails, as long as there are no mixes in the trial of 10 flips? What this means is that ending up with all 10 heads will work, and so will all 10 tails. Thus the probability of getting either outcome equals the sum of the separate probabilities of each outcome: $1/1024 + 1/1024 = 2/1024 = 1/512$. In other words, the result would be about 1 in 500 such tries, or double the more restrictive requirement of 10 heads in a row. Either of these mutually exclusive outcomes (10 heads *or* 10 tails) satisfies the condition of all coins showing 1 face in 10 flips. This means that 511 out of 512 (or 1,022 out of 1,024) instances when we might flip a coin 10 times in a row, we'll expect to get something other than all 1 face in those flips. We'd expect some mixture of heads and tails 99.8% of the time.

Suppose we require answers to more specific questions, such as what is the probability of getting 9 heads and 1 tail in a family of size 10 (a series of 10 consecutive flips)? Or 8 heads and 2 tails? Or 8 of either heads or tails and 2 of the opposite face? Or—more interesting, perhaps—how often would we see what we expect: namely, 5 heads and 5 tails? How do we answer such questions with statistical precision? That's where binomials come in.

Let's first apply it to the likelihood of 5 heads and 5 tails in 10 flips of the coin. We could crank out all possible outcomes using $(a + b)^n$ or target a specific outcome and ask what its likelihood is in the distribution of possibilities. The computing formula is $[n!/(s!t!)] \times [a^s b^t]$, where n is the size of the family (in this instance, 10); *s* is the number of times event *a* occurs and *t* is the number of times event *b* occurs (in this instance, 5 each); and *a* and *b* are the probabilities of the events "heads" and "tails" (in this instance, 0.5 each). Plug in the numbers, turn the handle, and we learn that the probability of getting exactly 5 heads and 5 tails in 10 flips of the coin (with the order of their occurrence not specified) is 252/1024 = 24.6%. That tells us that we see what we expect less often than what we don't expect. While this sounds like silly double talk, it means that some outcome other than half heads and half tails in 10 flips of the coin shows up, on average, about 75% of the time! What we don't expect to see (meaning something other than 5 heads and 5 tails) actually happens more often than not! How can this be? Is this somehow counterintuitive? Not really. The overall frequency of heads to tails over the whole distribution of possibilities of outcomes for families of size 10 is still exactly 0.5, or 50% each. That has not changed one iota. Of all the possible combinations of heads to tails in families of size 10, the exact division of an equal number of heads to tails (5 of each) is the most frequent outcome expected. But there are a number of other outcomes as well, and their exact probabilities can each be figured out mathematically. Their sum total exceeds the single most frequent occurrence. It's not a mystery. It's only when our observations differ significantly from our mathematical predictions that we suspect something other than chance may be operating.

I compared moth-settling behavior on black versus gray backgrounds with binomial predictions and discovered that their choices were not random. Something other than chance deviations from these predictions seemed to be operating. I presented the data for this and subsequent work, as well as the details of their analyses, in a lengthy research paper with a colleague, published in 1988 in the *Biological Journal of the Linnean Society*, so I won't repeat all that here. In a nutshell, the tails (extreme ends) of the observed distribution of black versus gray choices made by the moths vastly exceeded binomial predictions. While mixes of choices were expected and observed, the midranges of the distribution were significantly underrepresented, thus hinting at a behavioral polymorphism. The surprising part was that this polymorphism in background choices was not related to the genetically inherited melanic polymorphism phenotype. That is, *typica* (pale) moths were just as likely to settle and remain on black backgrounds as were the *carbonaria* (melanic) phenotypes, and, likewise, *carbonaria* were just as likely to settle and remain on gray backgrounds. Still, the moths were hardly indifferent to the reflectance of the backgrounds in the pen, as many of the same moths made identical choices repeatedly, day after day. But different moths made other choices, suggesting that individual variations existed among the moths, unrelated to their own inherited colors. Wow! Now this was totally unexpected. I had guessed that either the moths would be indifferent to the backgrounds and would settle randomly, or they would all show a statistical bias for the same background. What I'd hoped to see was a polymorphism in which the black moths settled on black, and the pale moths settled on gray (as Kettlewell had originally reported). What I did not expect to find was some moths settling on black and some on gray over and over again, with these choices having nothing to do with their inherited color, nor with the colors of the collars I'd put around some of their necks. There were, it seemed, individual differences in background preferences. If real, what did it mean?

22 | The Talk

The laboratory evidence for individual differences among peppered moths in their rest-site preferences teetered on a small sample of adults that had been imported as pupae from England. The progeny from some of these moths were, in turn, run through the rest-site selection pens the following year, to assess the heritability of that behavior. Unfortunately, too few of the crosses produced large broods. I had hoped to pair all-black-choosing males and females together and all-gray-choosing males and females together, along with reciprocal crosses of black and gray choosers. That would make perfect sense in any classical genetics study. Alas, peppered moths are not as cooperative in the lab as fruit flies are. As a former drosophilist myself, I was discombobulated.

To put this into perspective, let's talk about fruit flies a little bit more. Most people are broadly familiar with them, although a wide variety of tiny flies bear the same common moniker. The superstar among several thousand species in a single genus is *Drosophila melanogaster*. Since the time of Thomas Hunt Morgan, over a century ago, geneticists have studied this particular fruit fly intensively. It's still a major player. True, there is a respectable list of other *Drosophila* species that are also scrutinized, so technically one should be *specific* when referring to *Drosophila*. Yet most members of this humongous genus are not used in the lab. Why? Part of this is certainly based on history, with an extensive body of knowledge already firmly in place that can be built upon. Another reason the same handful of species are used repeatedly is convenience. Some *Drosophila* species are difficult to breed in captivity. Finding just the right food that will keep particular species going for generation after generation is one problem. Another is that some species seem unwilling to mate in captivity. Surely they know how to mate, but ordinary lab conditions fail to put them in the mood. The great drosophilist Herman T. Spieth called this "mating recalcitrance."

Peppered moths certainly are not shy about mating in captivity. When they do, they remain coupled throughout the following day, so when a mating does occur, there is no mistaking it. My colleagues and I have routinely kept broods of peppered moths in captivity over several successive generations. This has allowed us to use controlled crosses to work out the basic Mendelian inheritance of melanism, and such research has been confirmed repeatedly. Scoring the melanic phenotype is straightforward—one look is all that's needed. Individuals of desired phenotypes can be paired as they become available. If they mate and produce progeny, great. If not, no worries. More will come along to mate, sooner or later.

But with the behavior I was examining, I needed to put the moths through the test pen for several successive nights, to see if they made the same choices again and again, before I assigned them to a phenotypic category as a black-chooser, a gray-chooser, or indifferent (mixed choices). While some of these tested moths would later mate when put into a mating cage, it was pretty iffy as to which pairings were successful. With the evidence for the heritability of a background bias vague (at best) from these few crosses, my long-range plan was to plod along with a directional selection–breeding program for a few more generations, at one generation per year. But with the labor-intensive rearing of caterpillars this process involved, I was beginning to wonder if the game was worth the candle.

My shorter-term goal was to try a different approach: geographic variation in background reflectance bias. The 1986 moth season was a long one for me. After running the F_1 (first progeny generation) of my incipient selection experiment through the test pens during April and May, I left the care and feeding of *their* progeny (the F_2 caterpillars) to a trusted assistant and headed back to England to have another go at wild-caught material, using the newer test pen. By then I was much better acquainted with when adult peppered moths emerged at Caldy Common, so I planned on being there in June, the peak flying season. During this period I stayed with Win and Mick Cross, in their grand Victorian house only a short walk from Cyril Clarke's moth trap. It was a lovely setup for me, and I eagerly fell back into the happy routine of opening the moth trap each morning, often to Sir Cyril's greeting of "How many?" The arrangement also allowed me to get to know Mick much better, as we shared many a pint during our free evening hours at the Moby, an upscale pub in West Kirby named for Herman Melville's great white whale.

Mick used these occasions to probe me about my impression of the

Clarkes. As both the Crosses and the Clarkes were members of the West Kirby Sailing Club, Mick knew them via their common hobby of sailboat racing, and he also knew Cyril as Win's boss for many years. He described his wife's earnings as "pin money," much to her chagrin. Win was a highly capable, intelligent, and well-informed woman who was dedicated to her job—and to her boss. She understood the importance of his work and the centrality of her role in it. Over the years she learned all aspects of his butterfly and moth investigations, which involved species imported from all over the world. She and the other assistants had to keep them going in the greenhouses, with each species having its own diet and culture technique. She and her coworkers were invaluable to Cyril, but, no doubt, they were underpaid—and perhaps underappreciated. Still, there was no question of their loyalty to the boss and to his work. My guess is that Mick understood all of this, but his cheap shot that these women worked for pin money was not so much to disparage their wages, as, more likely, to belittle their boss. Cyril Clarke was notoriously frugal and operated his butterfly research on a shoestring budget, questioning every expenditure. Yet I suspected that wasn't Mick's real reason. He often grumbled that Cyril, as a fellow member of the sailing club, would come to weekend regattas, sail in his competitions (usually winning), and then immediately go home, without sharing in the camaraderie afterwards. Mick resented that Cyril would not hang around at the club to have drinks with the other sailors at the end of the day. Mick didn't specifically say that Cyril Clarke "thinks he's too good for us," but that was the implication of his complaint.

I attempted to disabuse Mick of that notion by explaining that Cyril had no time for small talk with anyone when he was busy. It had nothing to do with his feeling "too good" for his companions at the sailing club. He came there to do what he expected to do: *sail*. Once done with that, it was time to get back to work. Yet what needed to be accomplished on weekends? Well, everything normally done by his assistants during the week. Someone had to take care of the livestock, and when his assistants had the weekend off, Sir Cyril and Lady Clarke did all of that work by themselves. It was time consuming. "For Win to enjoy a bit of social life with you at the sailing club," I explained to Mick, "the Clarkes needed to do Win's work in the greenhouses."

I doubt my defense of Cyril Clarke left much of an impression, because I also suspect that Mick, while recognizing Sir Cyril's importance in the medical world, saw his butterfly investigations as merely an extravagant hobby. Hobby or not, whatever Cyril Clarke did, he did full speed ahead, with Lady Clarke right there with him, fully engaged, and that included winning at sailing. This is not at all to suggest that Cyril didn't enjoy a bit of social drinking with colleagues. He most certainly did, but not when there was work at hand. And there usually was. Yet after a long day, if I might still be on the premises, he'd boom out, "Bar's open. Have a big drink!"—an invitation hard to decline.

My own work, besides running the moth trap and tracking the behavior of marked individuals in my test pen, was spent analyzing the data from the F_1 generation scored in Williamsburg and making graphs to illustrate their behavior, in preparation for a talk I was to present at a meeting of lepidopterists organized by the British Natural History Museum. The meeting was to occur later in the month, at Kew Gardens. The organizers had invited me to give this talk on my work, although I strongly suspected that they had first invited Cyril, and he nominated me to speak in his place. I don't know that with certainty, but why else would I have been asked to talk about peppered moths to the leading lepidopterists in the United Kingdom before I had yet to publish one word on the subject? In any case, I was flattered by their request to speak at this august event and worked hard to prepare.

From June 1 through June 27, I had caught 332 peppered moths in Cyril's mercury vapor lamp and assembling (pheromone-based) traps, and I had run a fair proportion of them through my test pens. I also noticed a conspicuous drop in the percentage of *carbonaria* in the wild population since my last visit. Now it was time to go to London and unveil my findings to the professional world.

I found digs not far from Euston Station and worked out the route to Kew Gardens. This would require travel by both tube and bus.

On arriving at the symposium the next morning, I quietly gave my name to the receptionist, only to have it echoed loudly behind me: "Bruce Grant?"

I was immediately greeted with an extended hand and a smiling face.

"My name is Rory Howlett. I know all about your work."

"Really?" said I, puzzled. "How could you possibly know all about my work? I haven't given my speech yet."

"Cyril Clarke has written to everyone about it, so we know what you're up to," said Rory, still smiling and very cheerful. "We should talk."

"Now . . . before I give my paper?"

"Perhaps over lunch? You're not scheduled to speak until late this afternoon. I doubt I'll have much time to linger beyond your talk, but I came here expressly to hear it and won't leave before then. But, in the meantime, perhaps we might compare notes? I am doing exactly the same work you are doing."

(I should interject here that these are not actual quotes of what Rory said or what I replied. I am just attempting to recreate the gist of our conversation, as I recall it.)

"Exactly the same work?" I questioned. As it turned out, Rory was in the process of earning his PhD at the University of Cambridge, under the direction of Michael Majerus, and he was indeed investigating the rest-site selection behavior of peppered moths. When Majerus was tipped off by Cyril Clarke that I, too, was conducting research on this problem, that news became a matter of great urgency for Rory. A formal requirement for a PhD thesis is that it must be based on original work, not something previously published by others. Rory feared that he was about to be scooped, so of course he arranged for an invitation to hear my talk. This was also news to me. While I knew that speakers are invited to speak at prestigious symposia, I never before heard of the audience (called "delegates") being invited as well. This, it seemed, was a big deal! At least to me it was.

So, sure, I agreed to have lunch with Rory. It was very enjoyable, in fact. He gave me a manuscript he and Majerus had prepared for publication in the *Biological Journal of the Linnean Society*, offering preliminary investigations on where peppered moths rest in the wild, along with inordinate speculation on what it might mean. The real duplication of our efforts was not in this paper. Since we used different approaches to score background selections, there was little likelihood that Rory's thesis would be rejected on the grounds that his work had previously been published. Moreover, he had done nothing to test the contrast/conflict mechanism proposed by Kettlewell to explain how moths made choices. We both were relieved by our meeting.

We did wonder, however, why we were having such a hard time confirming Kettlewell's claim that melanic and pale peppered moths showed morph-specific differences in their rest-site selection. Almost as parting words, Rory cautioned me darkly not to call Kettlewell a fraud, as my audience would be packed with people who held him in high esteem. "Fraud?" I said. "I wouldn't dream of it. I'd call him a pioneer."

I did call him a pioneer, and I meant it. None of the research that was going on then would have been underway had it not been for Kettlewell, even if much of it aimed at reexamining his claims. That's how science works. Kettlewell himself was the first to find fault with his own methods and suggest improvements that his successors later explored. To paraphrase an old pearl: we stood on the shoulders of this giant.

Yet still I felt an outsider's isolation, at least at that stage, with my American foot jammed in the British door. I was hardly James Watson, trying to crack the structure of DNA, but perhaps I was invading their turf. Would they circle the wagons? What cheek we Yanks have! Don't we have our own moths to work on?

My talk went well enough, although the transparencies projected poorly in the summer sunlight. The room was not air-conditioned and was hot and stuffy by British standards, so the windows were opened and the blinds raised, to capture a breeze. It was a necessary tradeoff. I never know for sure how my talks go over with an audience. There is always applause for all such talks, as a matter of form. Nobody booed. From compliments offered to me afterward, I am willing to accept that my presentation was well received. One of the hosts from the Natural History Museum told me he never knew an American who didn't deliver a great talk, and I fully lived up to that tradition. I was tempted to suggest he hadn't yet heard from that many Americans but instead thanked him when he dangled an invitation that I return to talk about my *Nasonia* work. We left it at "maybe someday."

Cyril Clarke had returned to the meeting for post-session drinks after having skipped all of the afternoon talks, including mine. He had to attend the dedication of a new medical wing somewhere else in London. For him it was a command performance, as the Queen was to be there. One doesn't decline an invitation extended on behalf of Her Royal Highness. Now, back at the bar at Kew Gardens after the symposium was over, one of Cyril's chums told him it was a pity that he had missed my talk. As I was also belly up to the bar with them, I brushed this off, saying, "Cyril knew exactly what I planned to say, so he didn't miss a thing."

To this, Cyril gave his customary "Yes." Without missing a beat, Cyril's esteemed colleague stated, much to my surprise and pleasure, "You may have known what Bruce had planned to say, but you missed how very well he said it."

Cyril again said "Yes."

By this time I was starting to feel better about my talk. Of course, I am a sucker for praise of any kind. Who isn't? I didn't stick around long enough to milk the audience for more compliments, but I savored the nice things my fellow American, Lincoln Brower, said in his wrap-up of the session and thought about the only tough question I received at the conclusion of my presentation. John Turner, world-renowned for his extensive studies on *Heliconius* butterflies, was in the audience. Delivered by anyone else, his question might have seemed hostile, but he is the consummate professional. In effect, he said that barrel studies like mine and Kettlewell's, where moths are placed in a container, are such artificial environments that no one can argue that the moths are behaving normally. Ergo, one can essentially learn nothing about them that is relevant to their behavior in nature.

I had heard that argument before and was ready for it. In response, I said that no one would argue that barrel experiments model the real world, but we can still learn from them. By analogy, I suggested that one might discover differences in preferences for tea versus coffee between Americans and the British by tallying choices made by prisoners. People in jail are hardly in an ordinary human environment and probably do not behave normally in various ways. Maybe, given a choice, British prisoners would drink tea more often than incarcerated Americans, who generally prefer coffee. I went on to explain that while the moths in barrel experiments do not behave normally, and that no doubt there are other factors involved in rest-site selection that are not controlled in barrel experiments, we have learned that the moths are not indifferent to the surface reflectance differences (light vs. dark) offered to them. Their settling patterns are clearly not random.

Turner seemed to nod his acceptance of this restricted interpretation, and I had the data to back it up. It wasn't an argument. It was a conclusion.

With this in mind, I headed back to my hotel to pack up for my flight home. Between the bus leg of the trip and the transfer to the Underground ride to Euston, there was a grand Victorian pub on the corner. I decided to go in to have a pint and reflect on the day's events. British pubs! What a great contribution to culture.

23 | The Grand Pub

The huge, elegant Victorian pub occupied one corner of a busy intersection where the bus from Kew Gardens dropped off passengers to connect to the Underground. Most went directly about their business, but as mine was effectively done, except for the flight home the next morning, I headed straight into the pub to quaff some ale and reflect on the day's events. By this time, I was a seasoned pro and knew how to request a drink without inviting the bartender's questions about my national origins. Besides, this was London, so strangers were far too common to be remotely interesting.

I took a long pull on my pint and rotated my barstool to have a look at the upbeat crowd. I paid no special notice to any of the patrons, but my general impression was they were young, energetic, well dressed, and full of life and its promises. Maybe the moniker Yuppies wasn't then in use, and perhaps it never was in England. Not sure about that. But there seemed to be a room full of Yuppies participating in happy hour at a grand pub. I enjoyed their hopefulness, especially as I reflected on my own perceptions that my talk at the British Natural History Museum went well. I had had a lot of anxiety beforehand. I always do before every talk I give, no matter where or to whom. To do otherwise is to disrespect the audience. I owe my best to them, to myself, and, above all else, to science. There is no greater human enterprise, and I am a part of it, however small my part is. So there I sat, in that grand pub, drinking ale, pretentiously thinking about my humble but noble role in science, and looking at the happy throng of people drinking, talking, throwing darts, dancing and mingling. Then, out of the corner of my eye, I spotted a markedly unkempt woman standing in the doorway at one of the two major entrances into the tavern. Reality intruded. She was clearly casing the joint. But for what? I wondered.

She slid into the room like a wraith, knowing exactly where to go. She moved along tables where the occupants had temporarily moved away to throw darts, or go to the WC, or dance, or talk with others elsewhere, leaving their drinks unattended behind them. One by one, she swigged them down as she stealthily moved through a section of the room. She was like a bumblebee homing in on pollen-producing flowers, and smooth as an Olympic figure skater. Such maneuvers would give her a caloric and alcoholic fix. She could have grabbed a sandwich or two to get her vitamins, but personal health and fitness seemed to not be her immediate concern.

She'd nearly finished downing all of the assorted drinks—wines, cocktails, and whiskies, the glasses partial or full—along the length of a table when two burly bartenders brusquely arrived on the scene to drive her off the premises. They were physically imposing lads, with close-shaved heads and well-defined arm muscles. I'd have stopped whatever I was doing if they told me to. She snarled angrily at them and slipped away before they could get close to her. Gone. Out the door. Slick as silk.

Maybe the house gave those cheated out of their drinks a free round. I don't know. In whatever way that situation was resolved, the place was soon back to where it was when I came in. Yuppies enjoying happy hour. As my own thoughts drifted back to the day's events at the British Natural History Museum, out of the corner of my eye I caught sight of the very same woman, but this time at the other entrance to the grand pub, around the corner.

Just as before, she took in the lay of the land, but this time she did not delay before moving into action. She entered the room from the side opposite the one she'd used before and swigged down all the unattended drinks as quickly as she could before being rediscovered by the bartenders.

Of course it didn't take these chaps long to notice her activity, this time on the other side of the huge room, and once again they proceeded to drive her off. This probably wasn't anything new to them, and certainly it was not new to the pilferer. She knew exactly what she was doing, and she was good at it. The Yuppies seemed oblivious to all this and ordered another round.

As I followed suit, I thought about John Turner's question, and wondered anew at the usefulness of my work amidst a world of people with concerns so different from my own.

1 Bess Murray (*left*) and Jim Murray (*right*), 2006. Photo courtesy of Jim Murray

2 David West at Mountain Lake, Virginia, 1994. Photo courtesy of Jim Murray

3 Lady Féo Clarke (*left*) and Sir Cyril Clarke (*right*) at Caldy Road, West Kirby, England, 1984.

4 Sir Cyril Clarke's devoted assistants, posing with the author: (*from left to right*) Bruce Grant, Winifred "Win" Cross, Angela Urion, and Jean "Riki" Butler, 1984, West Kirby, England.

5 Angela Urion (*left*) and Sally Thompson (*right*), who was Win's successor, in the boss's study, 1999.

6 Leaving Liverpool, England. Piper Major Tom Graham (*far right*) and son Gordon (*far left*) piping off the Grant clan at Lime Street Station, 1984. The author (*center back*) with daughters Elspeth (age 9) and Megan (age 17).

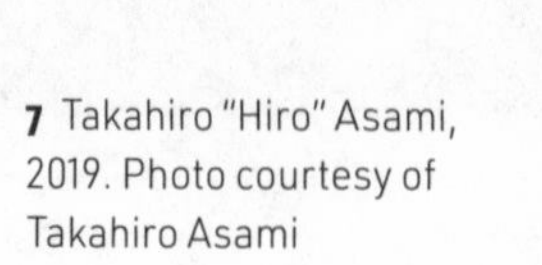

7 Takahiro "Hiro" Asami, 2019. Photo courtesy of Takahiro Asami

8 A portable Robinson mercury vapor lamp moth trap.

9 Dave West's not-so-portable mercury vapor lamp moth trap.

10 Denis Owen unpacking a moth trap on the Edwin S. George Reserve, Michigan, 1994.

11 Most moths are light inhibited and stay put on cardboard when removed from traps after sunrise. The largest moth shown here is representative of the *cognataria* caught at Mountain Lake, Virginia. Dave West's thumb (*lower right*) is included for scale.

12 A photo showing the extreme contrast between typical (*left*) and melanic (*right*) phenotypes of British peppered moths.

13 British typical (*lower arrow*) and melanic (*upper arrow*) peppered moths, posed by the author on a silver birch tree (*Betula pendula*), to illustrate that neither phenotype is particularly conspicuous when resting on an appropriately heterogeneous background.

14 A typical female (*upper left*) and a melanic male (*lower right*) from Michigan, posed together for comparison. At rest, their wings span 4–5 cm. Cover from Grant and Wiseman, 2002, reproduced with the permission of the American Genetics Association

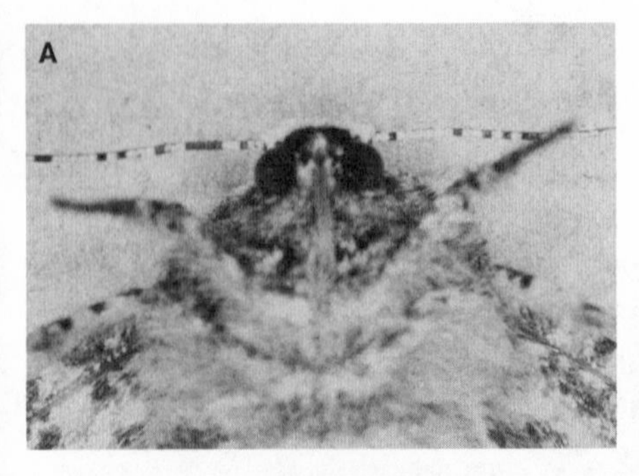

15 Uncollared typical (*A*) and melanic (*B*) moths, showing the natural scales around their eyes.

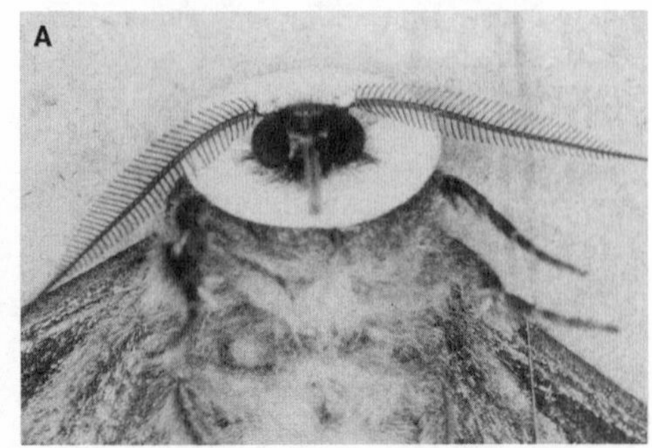

16 White (*A*) and dark (*B*) paper collars placed on peppered moths in attempts to fool them about their true colors.

17 A resting-background test pen.

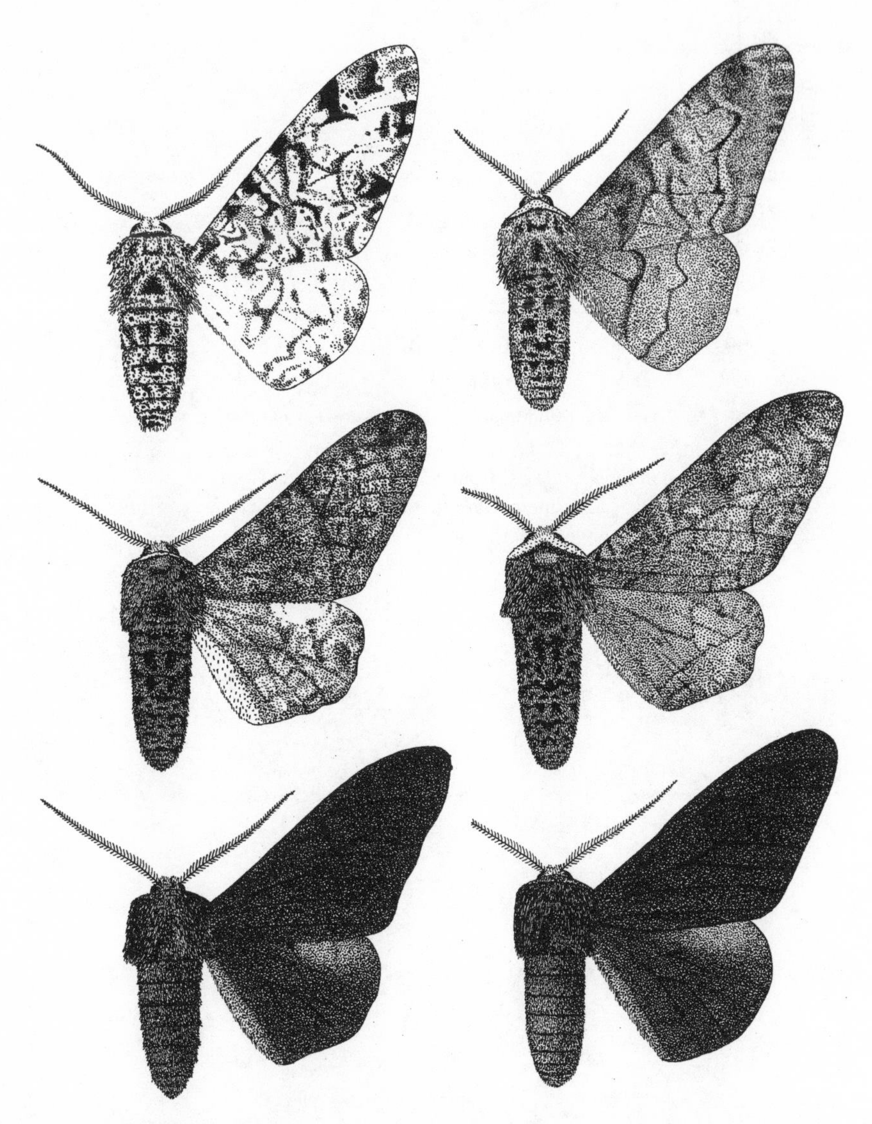

18 Derek Whiteley's side-by-side comparisons of wing variations and similarities between British (*left column*) and American (*right column*) peppered moths. Cover from Grant, Owen, and Clarke, 1996, reproduced with the permission of the American Genetics Association

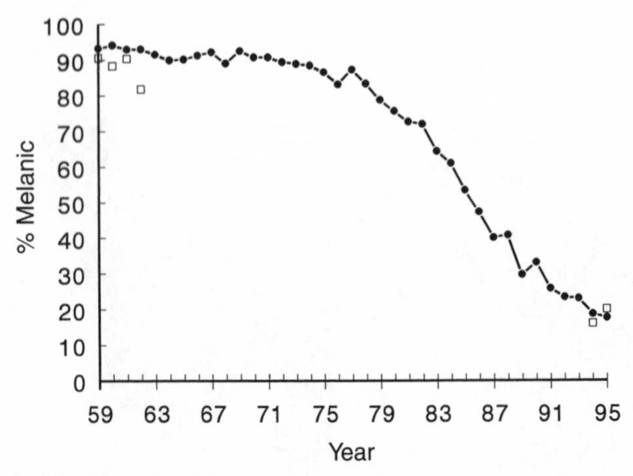

19 A graph of the decline in frequency of melanic *Biston betularia* at Caldy Common in England (*solid circles*) and at the Edwin S. George Reserve in Michigan (*open squares*) from 1959 to 1995. Adapted from Figure 1 in Grant, Owen, and Clarke, 1996, reproduced with the permission of the American Genetics Association

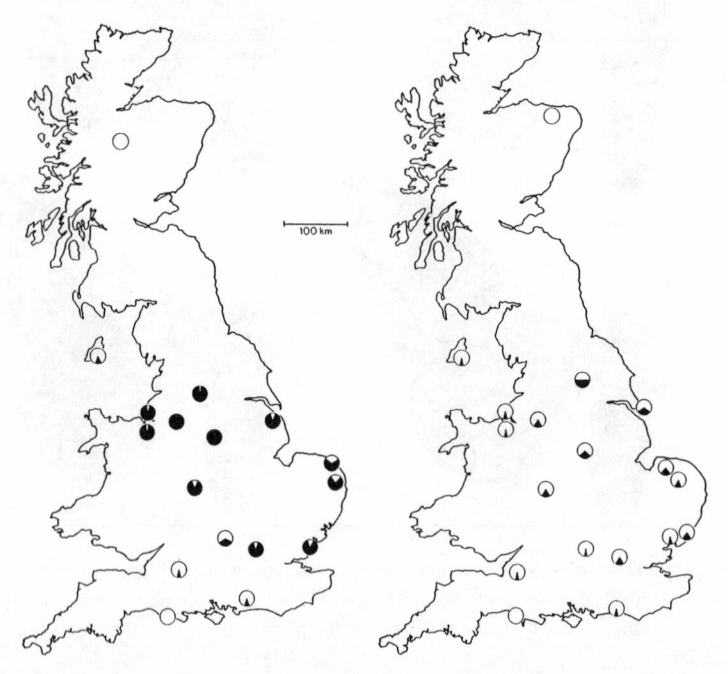

20 Before-and-after comparison maps of melanic frequencies across the United Kingdom between 1952–1956 (*left*) and in 1996 (*right*). The *pie charts* indicate the percentage of melanics at each location that was sampled. Figure 2 in Grant, Cook, Owen, and Clarke, 1998, reproduced with the permission of the American Genetics Association

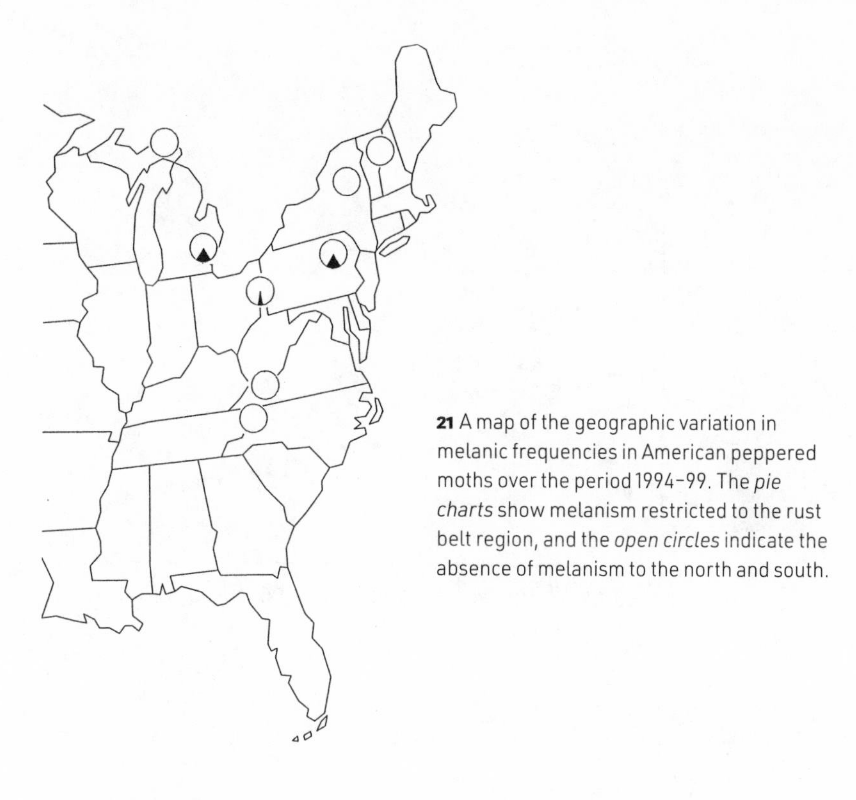

21 A map of the geographic variation in melanic frequencies in American peppered moths over the period 1994–99. The *pie charts* show melanism restricted to the rust belt region, and the *open circles* indicate the absence of melanism to the north and south.

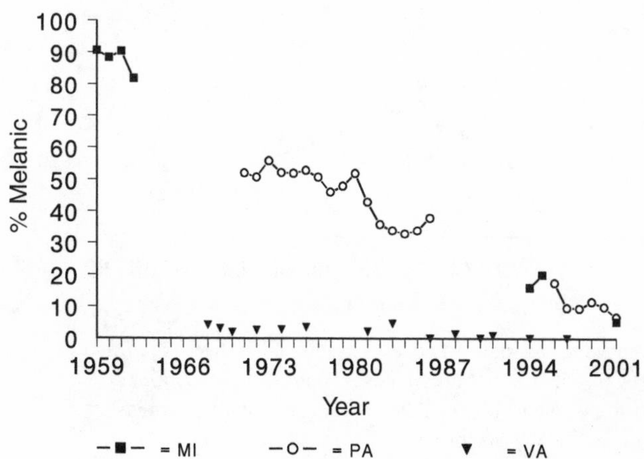

22 A graph showing the decline in melanic frequencies in American peppered moth populations at three locations: southeastern Michigan (*solid squares*), northeastern Pennsylvania (*open circles*), and southwestern Virginia (*inverted triangles*). Figure 1 in Grant and Wiseman, 2002, reproduced with the permission of the American Genetics Association

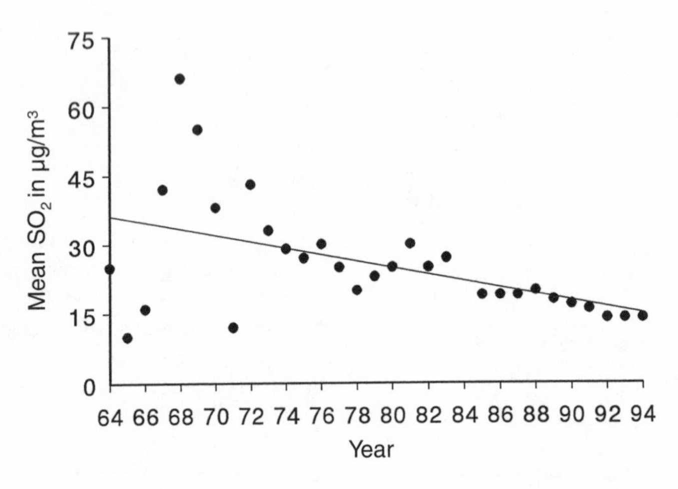

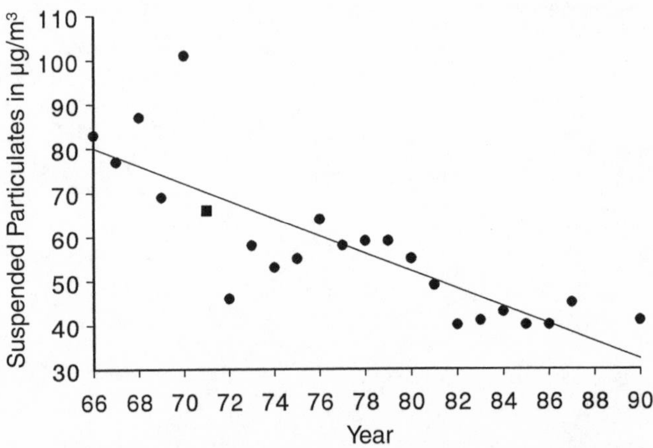

23 Graphs showing changes in air quality in southeastern Michigan, as measured by (*top*) SO_2 and (*bottom*) suspended particles (soot). Figure 3 in Grant, Owen, and Clarke, 1996, reproduced with the permission of the American Genetics Association

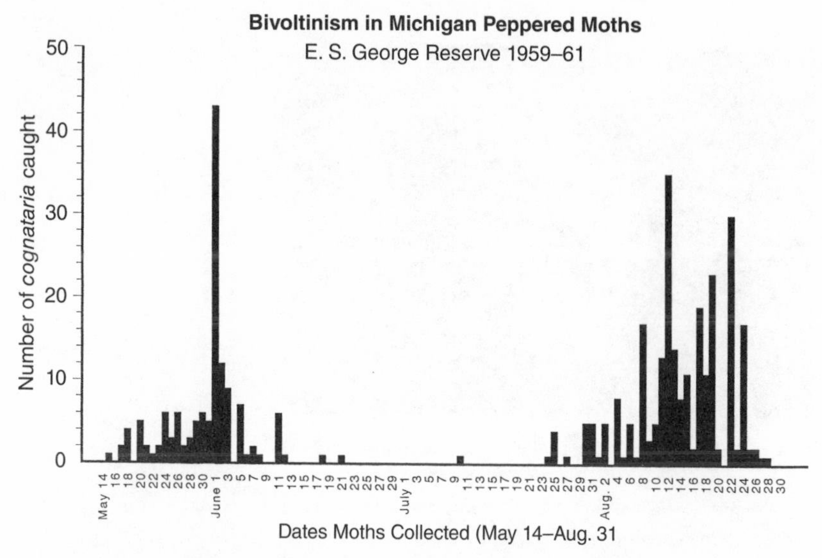

24 A graph showing bivoltinism in American peppered moths collected at the Edwin S. George Reserve in southeastern Michigan from 1959 to 1961. There are two generations, or broods, of adult moths each summer, with very few adults on the wing in June and July. The author constructed this graph from the dates shown on the tags of pinned specimens deposited at the University of Michigan's Museum of Zoology by Denis Owen

25 Caterpillars develop darker (*left*) or lighter (*right*) colors that closely match the twigs of their host plants, unlike the colors of adult moths, which are determined by genes and do not change. Figure 1 in Noor, Parnell, and Grant, 2008

26 The William & Mary caterpillar crew, 1992. Robin Parnell (*left*) determined that vision stimulates caterpillar color development, and Mohamed Noor (*right*) demonstrated that color development is reversible (see Noor, Parnell, and Grant, 2008). Photo by Bruce Grant

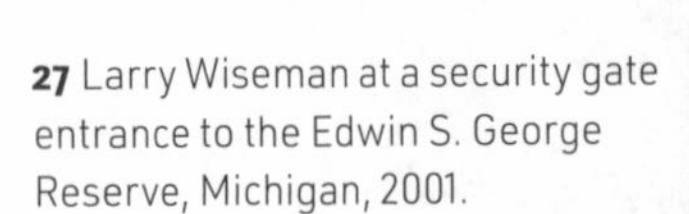

27 Larry Wiseman at a security gate entrance to the Edwin S. George Reserve, Michigan, 2001.

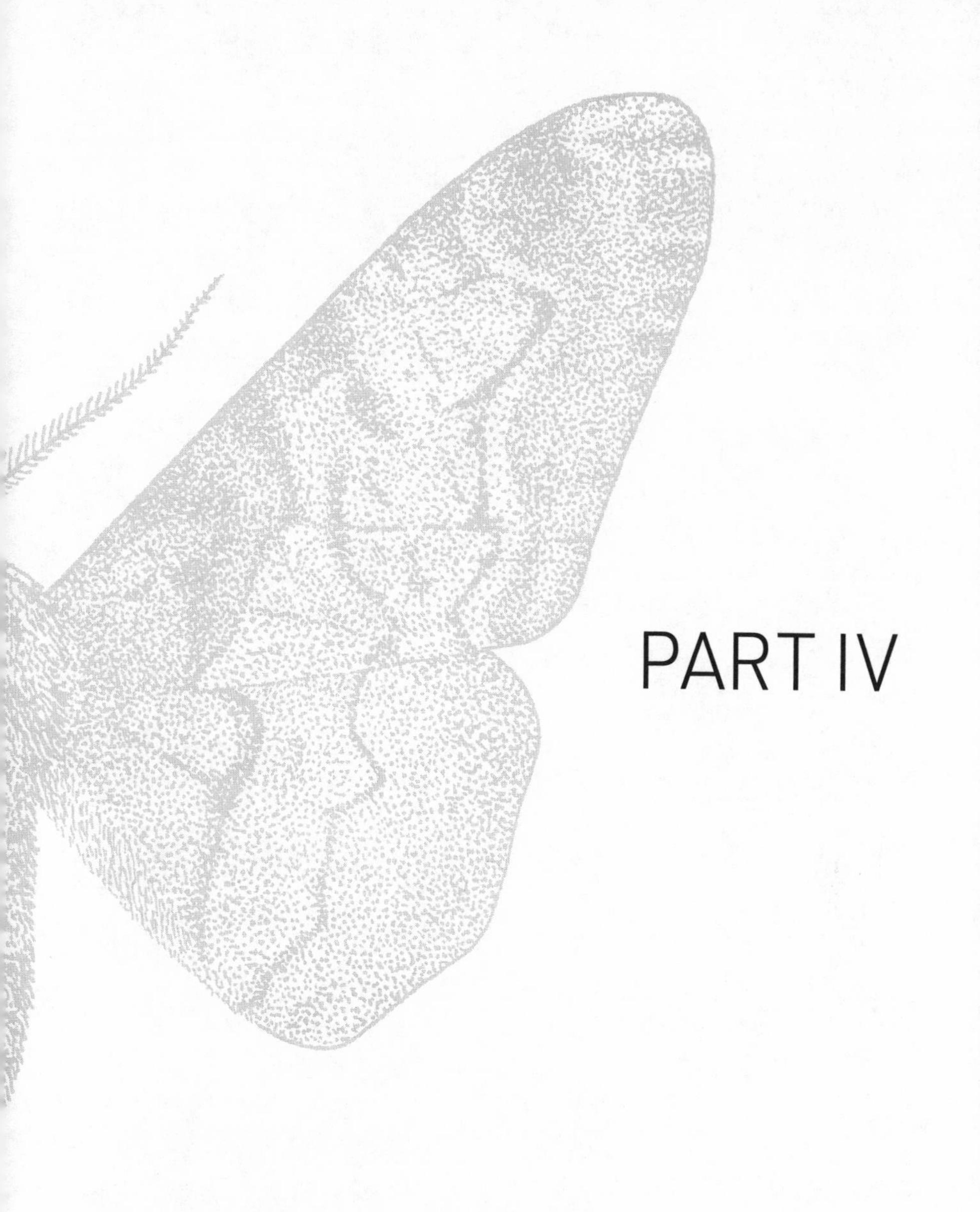

PART IV

24 | Summer School

Latitude made it possible for me to milk two peppered moth seasons out of a single summer. In the United Kingdom, peppered moths fly from the middle of June, throughout July, and are finished by August. At more southerly latitudes, such as at Mountain Lake in Virginia, peppered moth adults are on the wing in the early spring and then again in August. I'll have more to say about this relevant detail later on in my saga, but I will state here that I was able to catch peppered moths in England during the first half of summer, and in Virginia during the second half of summer, in the very same year. How wonderful is that?

Upon returning from my moth season at Caldy Common, running a moth trap in Cyril Clarke's garden, I was once again at the Mountain Lake Biological Station in southwestern Virginia, operating a moth trap just days after having given a talk on my barrel experiments at a symposium at Kew Gardens in London. I had become a jet-setter scientist! More pedestrianly, I had been invited to teach a course at Mountain Lake and was flattered to do so.

Normally I avoided teaching in the summer. During the back-to-back semesters of the regular academic year, I poured my full energy into preparing and presenting lectures. While I enormously loved doing it, I needed summers to recharge. I didn't want to risk burnout. So in summer I preferred to focus exclusively on research. Then, come fall, I very much looked forward to the start of the new academic year and was always excited to get back into the classroom. I made an exception to my rule by accepting the teaching job at Mountain Lake, because I wanted another crack at its peppered moths. I could have run traps there again, even without teaching, as I had done initially, but I opted for the full treatment. It turned out to be a full-time job—total immersion! Still, I enjoyed it. The students were terrific. I think we all had a good time and everyone learned a lot. A few years later, I did it again, but after that I declined further invitations to teach summer school.

The course itself was essentially my version of Dave West's long-standing Ecological Genetics, but as I was not a *field* biologist, I changed the title to suit my own focus: Evolutionary Genetics. I was quite comfortable with the lectures. That's what I did for a living and enjoyed high ratings throughout my career. It was a new challenge to figure out how, in a five-week summer course, I could squeeze in hands-on labs to demonstrate evolutionary events occurring over generations using real, living organisms, rather than simulations. Microorganisms could do the trick. But this was a field station, where people spent a lot of time outdoors, in the woods, looking at critters. So I relied heavily on *Drosophila*, my first love, to illustrate evolutionary processes that could be directly observed in the laboratory. Getting with the program, my students placed *Drosophila* population cages on windowsills, so the flies could see the field outside. Nature!

The problem for me, as a lab experimentalist teaching at a field station, was to take the students into the actual *field*. So I played to my strengths, narrow as they were. We ran flytraps outdoors, and I taught the students how to use a taxonomic key to identify fly species. Pretty good. No one else at the station knew how to do that. I did. I also taught them how to trap moths and identify them. Not many at the station then knew how to do that, either. But mostly my projects for the students were experimental laboratory genetics. As a field biologist, I was a novice myself. I felt no need to apologize for this. Testing evolutionary theory in the laboratory was what I did. L. E. "Gene" Mettler, my incredibly talented mentor, called our specialty "experimental evolution." I was thoroughly at home in this discipline, but at Mountain Lake that summer, I was being paid to teach a graduate course at a field station, a somewhat different kind of biology than I was used to.

To flesh out my otherwise meager offerings, I imposed on a few resident experts, who generously donated their time to present labs or organize field trips. On one outing, we spent the afternoon sampling a population of land snails (*Cepaea nemoralis*) that had been accidentally introduced to North America from Europe. Variations in shell banding patterns within and among populations were explained as products of differential predation by birds on snails living in different habitats, or the result of genetic drift, or some combination of the two. It was an old textbook classic, so finding *Cepaea* here in Virginia was a real kick. Jim Murray showed us where the colony was located. Years later I discovered another colony in Canada, and I collected a sample of snails from it to give to Jim as payback. In any case, we spent a long afternoon in the field. Meanwhile, back at the main building, a note marked "URGENT" indicated that I had a phone message. Everyone who walked past the bulletin board that day read it.

At that time, the station only had two phones: one in the main office and another in a back room. These were answered either by the secretary (during office hours) or by anyone who happened to pass by when a call came in. There was no practical way to track people down. No one carried cell phones back in the '80s. So notes about calls were posted on the central bulletin board. The person who had phoned me that day was a fellow piper in the band I played with in Williamsburg. In fact, I taught him to play the bagpipe when he was still a high school student. He later went on to the University of Virginia (UVA) and ultimately became a lawyer. As he knew I was a geneticist by profession, he hit me up from time to time for expert advice. Some of the cases he handled involved disputed paternities. Early on in his career, A, B, O, and Rh blood-type evidence was used in court. Then he moved on to HLA (human leukocyte antigens), before getting into more-recent DNA fingerprinting. So he sometimes wanted to run the evidence he had in specific cases by me, to find out whether an accused party could be excluded. While I was away snail hunting with my class, he called me to discuss a case. His urgent message, posted for all to see, read: "Bruce, call your lawyer about paternity suit."

Biology, like physics, has its own pecking order among its subdivisions. Those attracted to various specialties are, perhaps unfairly, stereotyped. I am willing and able to make fun at the expense of my colleagues at beer-drinking sessions, but I choose not to put this silly sport into writing here. Ranking these subdivisions in some sort of hierarchy is senseless. In every specialty there are dedicated and talented people who could just as easily excel in any discipline they chose to pursue. Indeed, a brief review of the history of some of the greatest biologists reveals people who have changed fields several times, not to find something easier or harder or more prestigious to study, but simply to follow their interests. George Beadle comes to mind. I am hardly in his league. Few are. I went into genetics, not because this is the discipline at the top of the pecking order in biology (according to some geneticists I could mention), but because, at the time I decided on that field as my career choice, it was the most interesting one to me—and still is. My abrupt shift toward field biology, perhaps regarded as being at the bottom of the totem pole (also according to some geneticists I could mention), was not forced upon me by teaching at a field station, but by my very specific interest in testing natural selection directly—*evolution in action*. To do my work, I had to get to know peppered moths personally. I had to become a field biologist. Indeed, that's where the real biology happens: in the field!

But woe is me! At field stations there is also a pecking order, determined by correctly identifying species at a glance. This is especially true for birders. Among these folks, I am at the bottom of their pecking order. Still, I very much enjoy being a bird *watcher*. I do it every day. Nonetheless, I especially admire those who can turn over a rock and identify everything they see. As I mentioned previously, few are better at this than Jim Murray, who also happens to be an evolutionary geneticist—easily at home at either end of the totem pole.

As it turned out, Jim's student Hiro Asami was another multitalented individual. After a few years as a UVA graduate student and a regular Mountain Laker, Hiro also blossomed socially. He gave one of the Sunday night entertainment talks while decked out like a samurai, minus the sword. He took a break from snails to describe in detail how he collected moths in his native Japan. Hiro's humor had us rolling in the aisles. We learned a lot, too. For one thing, he didn't use a Robinson trap to catch moths. That method was well suited for population studies, but for fussy moth buffs who sought perfect specimens to spread and pin for display in their collections, Robinson traps can be a bit rough on the captives. Hiro, like professional lepidopterists, used a white bedsheet stretched over a clothesline. The sheet is illuminated by a bright light (usually with a mercury vapor bulb) shining on it, similar to a movie screen at a drive-in theatre. Moths land on the sheet and soon settle in place, without losing too many scales in the process. The collector then lifts them gently from the sheet and puts them into a container (usually a "kill jar"). As I watched Hiro's slide show, I suddenly noticed, among the several dozens of moths shown clinging to a collection sheet, a familiar delta-shaped moth clamped to the lower right corner. Unmistakable! Nothing else looks quite like a peppered moth, although lots of geometrids are similar. By this time I had become an expert, after having a few seasons of moth hunting under my belt. I raised my hand and called out, "Hiro, is that a peppered moth in the lower right corner?"

He then carefully examined that part of the screen and said, "Yes, it is."

"Wow!" I exclaimed. "So, you have peppered moths in Japan?"

"Yes."

I pressed on: "Are there melanic forms of peppered moths there?"

To my disappointment, Hiro said, "I have never heard any reports of melanism occurring in Japanese peppered moths."

"Hmm," I mused. I found that curious. Japan was the Asian counterpart of the United Kingdom, being heavily industrialized, with its own subspecies of peppered moths (*Biston betularia parva*), so it seemed like a logical place to look for parallel evolution. Replicated experiments are mandatory in laboratory routines, but Japan might offer a golden opportunity to find a rare replication provided by nature. As Hiro was finishing his talk, I made a decision then and there: I must go to Japan to have a look. To do that, I'd need Hiro's help.

25 | Coauthors

While I was still teaching at Mountain Lake, I received a letter from Rory Howlett, the then University of Cambridge graduate student I'd met at Kew Gardens earlier that summer. In it, he suggested that we collaborate on a paper. As he put it, we were bucking the establishment, and by joining forces, we had a better chance to be heard. Made sense to me.

My next meeting with Rory was in January, at the Richmond airport. From there I drove him to Williamsburg, where he stayed at my house while we spent the next several weeks writing our joint paper, entitled "Background selection by the peppered moth (*Biston betularia* Linn.): individual differences," published in the *Biological Journal of the Linnean Society* in 1988. I was the senior author, not only for alphabetical ordering, but also because, by far, most of the data presented were mine. Still, it was a productive collaboration. While I have written papers with other coauthors before and since, this one really was written *together*.

My Evolutionary Genetics course lectures at William & Mary were delivered in the mornings, while Rory slept at my house. I'd then go fetch him for long afternoons of collaboration, which were spent on data analysis and writing. We composed the paper in the departmental computer room. During the 1980s, we did not yet have the luxury of individual desktop computers in our separate offices. That came a few years later, in the early '90s. Today it's hard to believe how we got along without our very own personal computers, but when Rory and I wrote our joint paper, we sat side by side in a communal computer room on the third floor of the biology building. Rarely was anyone else there, except for Charlotte Mangum (the most globally recognized biologist at William & Mary at that time), who seemed to always be there. She did have her own machine, but she kept it in the common computing room, rather than in her cramped office. She also had her own special computing chair. No one else would dare sit in it, even when Charlotte was away. The faculty computing room on the third floor was, in reality, an extension of Charlotte's office. Rory and I made our way there each afternoon to work on our paper.

Years later, Charlotte would comment, with much amusement in her voice, about what a strange working relationship Rory and I had. "I'd never seen anything like it," she said, "like I'd imagine Lennon and McCartney did, sitting at a piano tossing out lyrics."

Rory would say "Yes," not unlike Cyril Clarke's affirmations. I would type away, just as I'm doing now, with two untrained fingers, as Rory stared at the screen. Words appeared and went away, or stalled to his "Yes" or "No" or "Hmm." Over the course of my career, I've published roughly 40 papers, and about half of them have included coauthors, but none of those have been so closely cowritten. That was in 1988. We hoped the paper would attract attention.

Rory later told me he thought we'd left things up in the air. I'm not sure I agree. While trying to determine whether Kettlewell's explanation for background selection in peppered moths occurring through self-examination—his contrast/conflict model—or Sargent's genetically determined and unmodifiable mechanism was correct, we discovered, through extensive experiments, that there was no demonstrable morph-specific behavior that needed to be explained. That was indeed real progress, not a failure to find an explanation for something that did not, in fact, exist.

What *did* exist—supported by an overwhelming abundance of evidence from many sources—were the rapid changes in the frequencies of melanic phenotypes within regions, and the striking differences in geographically isolated populations of peppered moths. The cause of these temporal and geographic changes remained the central questions and required the closest scrutiny. Rory and I both fully understood this, but at the time of our collaboration, Kettlewell's versus Sargent 's explanations for background selection distracted us. Still, our time wasn't wasted. We managed to put to bed the idea that morph-specific background-selection behavior was an active ingredient in the evolution of melanism in peppered moths.

We enjoyed our collaboration and, eventually, each other's company. As anyone might guess, an Oxford graduate with a Cambridge PhD must be a very bright person. Rory thoroughly filled that bill. At first our days spent together were awkward and stiff, but as we got to know each other, we clicked rather nicely.

Writing the paper in close collaboration developed from being initially excruciating to tolerable to finally rather efficient. At the end of our long afternoons at the computer, Rory and I would go to Paul's Deli for a pitcher or two of beer before heading home for supper. This was before the era of fashionable craft beers, so we drank Budweiser, brewed right there in Williamsburg. Rory pronounced it "Budweezer," for some reason. Maybe it was his Oxbridge diction?

Rory was also an occasional cigarette smoker, though he never bought his own that I can recall. On one of our excursions to Paul's, when I returned to the bar with an empty pitcher for a refill, surrounded there by William & Mary students and faculty enjoying happy hour, Rory called after me from our table, in his booming Oxbridge accent, "Oh Bwuce, pick up some fags." This jarring word is familiar slang for cigarettes in the United Kingdom, bringing to my mind, not for the first time, Bernard Shaw's observation that the Americans and British are two peoples separated by a common language.

To many, our pub sessions might seem like an imprudent use of time, but we had our most productive conversations over beer. We didn't talk about football (theirs or ours), nor did we discuss movies, actors, politics, money, or women. We conversed about our common passion: peppered moths.

I recall an open-house reception at Mountain Lake the previous summer, when a well-educated woman asked me what type of research I did. I told her I studied the evolution of melanism in peppered moths. She replied that I was lucky to find someone to pay me to do that.

"Lucky?" I said. "I'd probably do it even if no one paid me."

"How lucky you are, indeed," she repeated, as if I were a hobbyist getting a salary to, say, collect stamps. She focused on "peppered moths" rather than on the word "evolution."

So I continued our conversation, defining what I did and explaining why I deserved to get paid: "As an example, what if a researcher says he's testing potential carcinogens on lab mice. Would you think he is especially concerned about the well-being of *mice*? Perhaps you'd recognize that he is interested in cancer and uses mice to see what agents might cause it. My interest," I went on, "is in *evolution*, and the evolution of melanism in peppered moths allows us to directly study how this biological process, occurring through natural selection, works in the real world, since we can witness it as it actually happens in nature. We must understand such things, not only to explain our surroundings, but to realize how, for example, superpathogenic (highly resistant, disease-causing) bacteria have evolved in recent years through the indiscriminant use of antibiotics in medicine and animal husbandry."

"What I do is not a hobby," I persisted. "I am not a moth collector. Not that there's anything wrong with that," I added, for a Jerry Seinfeld effect to lighten the pomposity, but this was too little, too late, as I continued to hammer away. "I'm a scientist who observes the process of evolution as it occurs, not as a facet of history."

"Oh," she said.

Whenever I bother to state all of this, I usually meet with different responses, ranging from "Aren't you lucky to have someone support your hobby" to "Yes, we should pay attention to what you do." Yet I rarely ever take the time to explain *why* I do what I do. In truth, it's because I enjoy it. Still, if you want people to pick up the tab for your efforts, you must supply reasons for why what you do matters to *them*. What do they get out of it? Why should they pay for your hobby? Fair enough! For me, it's easy. Understanding how evolution works is crucial to our comprehension of what we are doing to our planet through any and all forms of environmental modification, including the use of antibiotics, pesticides, genetically modified organisms, industrial pollution, and overpopulation. I should probably say that I am interested in how environmental modifications effect changes in the genetic makeup of populations. People might be willing to pay me to do that. But to say I am interested in the evolution of melanism in peppered moths makes me sound like a hobbyist—similar to an amateur bird watcher.

Rory is a bird watcher, too. No one pays him to do that. So, besides drinking Budweezer at Paul's, we also went birding. I took him to the Surry Nuclear Power Station at Hog Island, just across the James River from Jamestown Island, where astonishing numbers of migrating birds gather during the winter in the cooling ponds for the reactor. Rory had never seen anything like it. Nor had I. It has since become one of my favorite places to show to visitors. Because many wonderful books have been written about birds by people far more knowledgeable than I, I shall spare you my inexpert descriptions of the sights we beheld there, other than to say we had a Big Year!

Whether we watched birds, played chess, or strummed guitars, we also had to confront the uncomfortable fact that we had failed to corroborate Kettlewell's morph-specific polymorphism. Confirming either Kettlewell's contrast/conflict model or Sargent's genetically fixed behavior for background preferences would have been fully satisfying, but what we came up with is the absence of a clear-cut, demonstrable behavioral polymorphism linked nonrandomly to the unambiguous, genetically inherited, qualitatively distinct phenotypic differences between melanic and typical peppered moths. We had a hint, through the statistically significant individual differences within local populations, that genes were involved in the background-selection behavior, but these were independent of inherited color polymorphism.

As color polymorphism was *transitional*, en route from the monomorphism of one form to the monomorphism for the other, the presence of both color phenotypes together, in appreciable numbers, was merely temporary. Thus coadaptations of independently inherited characters (traits) through genetic expression, or linkage, seemed highly unlikely. Kettlewell's contrast/conflict model was far more attractive, because, as a single-gene pleiotropy, it required no time to evolve. Yet our experiments failed to support it.

This gave us plenty to think and talk about. One nagging possibility was that Kettlewell made it all up, and that we were dupes on a snipe hunt. Was Kettlewell a fraud? How else to explain the gross disparity in our data? He reported a morph-specific polymorphism in background selection behavior. We failed utterly to corroborate even that, let alone contrast/conflict. Why?

His detractors might all too easily call him a fraud. I myself had briefly entertained such thoughts. But over my many years in this business, having met and gotten to personally know so many honorable people who have worked side by side with Kettlewell, eyeball to eyeball, I have come to learn that all of them—every single one—regarded him as incapable of committing fraud. Yet they all also regarded him as a forceful personality, full of himself and his opinions. He was apparently quick to jump to conclusions and interpretations, kept no field notes, and was impatient with fussing about details. But overt fraud? Never! No one who knew him saw that as even remotely possible. He might arrive at wrong conclusions through stubbornness, but not by dishonesty.

Well, there we were, Rory and I, drinking Budweezer at Paul's, wondering why our barrel experiments and Kettlewell's produced such different results. "OK, let's accept that he did not fabricate his data," one of us might have said, "and concede that his published data are real."

"Of course they are real! Here's why. *Geography*!"

What Kettlewell did in his barrel experiments very likely followed the same pattern regarding the *source* of the moth subjects he used in his predation experiments. In the latter studies, he released *equal* numbers of typical and melanic peppered moths into woodlands in different parts of England. In Dorset, where light-barked trees were covered with lichens, essentially the only moth phenotypes available there then were typicals. Where would he get melanics to release in Dorset for his experiments? Moreover, well to the north, near Birmingham, where trees were blackened by soot and devoid of epiphytes, the melanic phenotypes had all but replaced *typica* as the common form. So, to have an available supply of equal numbers of both phenotypes (melanics and typicals) in both habitats (polluted and unpolluted woodlands) for his experiments, Kettlewell established his sources for the two phenotypes from two different parts of Britain. He was criticized for this practice, as he was introducing a potentially confounding effect into his predation experiments. Had he likewise employed typical and melanic peppered moths derived originally from different populations, his barrel experiments would have been seriously confounded, and the morph-specific polymorphism he reported could easily be explained by *polytypism* (the genetic differentiation between geographically separated populations of the same species that are adapted to local conditions). In other words, in the unpolluted woodlands of Dorset, directional selection not only clearly favored the light-colored typical phenotype, it also favored a behavioral bias for light-colored rest-site backgrounds. In the heavily polluted woodlands of Birmingham, directional selection instead favored the black (melanic) phenotypes, especially those among them that actively selected dark backgrounds on which to hide from predators.

Because Kettlewell left no field notes, we may never learn whether our proposed explanation is true, but if it is, there is a certain irony here. Kettlewell legitimately criticized Sargent for missing the fundamental distinction between (1) polymorphism within populations of the same species, and (2) variations among monomorphic populations of different species with dissimilar evolutionary histories. The very same criticism would apply to Kettlewell's experimental design if the typical and melanic peppered moths he used were derived from different populations. Had Kettlewell fooled himself?

Years later, on a trip to Cambridge to visit Michael Majerus, Rory's former PhD mentor, I essentially repeated our conversation, but this time with Mike. We covered the same ground and came to the same tentative conclusion. Until we learn otherwise from, say, the discovery of Kettlewell's field notes or other evidence, we feel reasonably certain that his barrel experiments were confounded by using typical and melanic peppered moths originally derived from different, geographically isolated populations. This does not mean the gene flow between them was zero, but it was sufficiently reduced so that significant genetic subdivisions would be expected under the varying selective pressures imposed by environmental modifications associated with industrial and urban development.

The paper Rory and I published in 1988 clearly showed polytypic behavioral differences among geographically isolated populations (British and North American) of peppered moths, and that is completely consistent with our conjecture about Kettlewell's disparate barrel results. For years afterward, I continued to use peppered moths from various geographic locations in my own barrel apparatus and consistently found no evidence supporting Kettlewell's report of a morph-specific behavioral polymorphism in background preferences. We did not leave that idea up in the air. We shot it down.

PART V

26 | *Nihongo*

The ad read, "Japanese lessons for sale."

"Looks authentic," Karen commented as she handed me the classified section of a local newspaper. Karen was a William & Mary undergraduate, working in my lab on research for her senior honors thesis. She was a straight A student who had been accepted into a PhD program at Princeton, but later opted for medical school at UVA. She knew of my plans to spend an upcoming sabbatical as a visiting professor at *Toritsudaigaku* (*Toritsu*, for short), or Tokyo Metropolitan University. This arrangement was made through the good efforts of Hiro Asami.

Hiro was a highly regarded graduate of *Toritsu* and stayed in touch with several of his former professors there, whom he persuaded to host my visit. They even agreed to orchestrate my travels around the country in search of *Oo-shimofuri-eda-shaku* (meaning "frosted, branch-measuring moth"), aka *Biston betularia parva*, the Japanese peppered moth subspecies. For obviously selfish reasons, I'd hoped Hiro would go with me. As I envisioned it, he would drive and do all the talking, while I focused on moths. Easy! Alas, Hiro hadn't yet pulled together all of the final pieces for his PhD thesis and had to stay in Virginia for another year to finish his degree. That meant I was to go to Japan alone and do all of the fieldwork by myself, without the benefit of an interpreter.

Karen's "Why don't you buy yourself some Japanese lessons?" prompted me to pick up the phone (at William & Mary, I had one in my office) and ring the instructor.

Her name was Shizuko Skoglund. She tutored the children of Japanese families who worked at the local division of Canon Inc., so the children wouldn't fall too far behind in their studies by being away from their native country for extended periods. Mrs. Skoglund also excelled at the traditional art forms of her culture, especially origami. She was a recognized professional and her elaborately folded paper creations were sold in gift shops at the Smithsonian Institution. I knew I had the right *sensei* to teach me Japanese (*Nihongo*) from the minute she graciously welcomed me into her house with *doozo o-hairi kudasai* for our first lesson.

Two other people were already there when I arrived. One of them didn't return after our first week. The other, Aileen, stayed for the whole several-week course, and then we signed on for a second round. Aileen's name might have been Irene, but I wasn't sure from Mrs. Skoglund's pronunciation. I was too embarrassed to ask, so I also called her Aileen. She didn't correct me, but maybe she was just being polite? Aileen had a PhD in engineering or physics and worked as a rocket scientist for NASA at their Langley, Virginia, research center. Her work required occasional travel to Japan; thus her interest in learning Japanese. Aileen and I were both highly motivated students, and Mrs. Skoglund poured as much Japanese into us as she could manage in the equivalent of a semester.

Toward the end of our second series of lessons, we enjoyed a traditional Japanese meal, prepared and beautifully presented by Mrs. Skoglund, and were tutored on how to use *o-hashi* ("chopsticks") properly. Through all of this, we three became friends, and over the years we had several class reunions, usually at Japanese restaurants, where I enjoyed showing off how proficient I'd become in using *o-hashi*—and *Nihongo*! We learned that Mrs. Skoglund was the widow of an American serviceman who was stationed in Japan during its occupation after World War II. They moved to the States and had two daughters. My wife played the organ at the wedding of one of the daughters, held at William & Mary's Wren Chapel.

Mrs. Skoglund's careful tutoring served me well during my sabbatical in Japan, and I hereby absolve her of any responsibility for the frequent miscommunications I committed as I wielded *Nihongo* like a *gaijin* (outsider).

27 | *Gaijin*

I left Williamsburg so early in the morning on May 29, 1988, that I had breakfast at John F. Kennedy Airport at 6:23 AM. This part of the trip included a flight from Norfolk to LaGuardia Airport and a taxi ride from there to JFK. As luggage I carried two soft-sided duffle bags, crammed full of what I thought I might need in Japan, plus two carefully wrapped glass bottles of Johnny Walker Black Label. The Scotch I brought as a gift for my host, Professor Osamu Kitagawa, the head of a university genetics lab in Tokyo where I'd be headquartered, on and off, for the next four months. Hiro recommended that particular brand as Kitagawa's personal favorite. Suntory was already producing Japanese Scotch, but Kitagawa preferred imported stuff for some reason, not that I could tell the difference.

By the time we landed in Narita, the major international airport serving Tokyo, I was pretty punchy after flying halfway around the world. Of course, the pilots and crew did all the work. I just enjoyed free drinks and slept. International travel can be so taxing!

I was surprised by the level of security upon our arrival. By today's standards it'd seem routine, but I had not previously seen such military oversight at commercial airports in Western countries. I just shuffled along like everyone else and ultimately boarded a shuttle to Tokyo's city air terminal. I was eager to see what Japan looked like, but that trip shot along a route enshrouded by high concrete walls, now all too familiar along urban thoroughfares. Might as well have been in a tunnel.

From the air terminal, I took a taxi to my digs in Minato-ku. Tokyo is a sprawling metropolis, with its central heart surrounded by formerly separate cities that have been incorporated into it, not unlike New York City, which, strictly speaking, is Manhattan, while the metropolitan Big Apple also includes several boroughs, such as the Bronx and Brooklyn.

For the first week or so I stayed at the International House of Japan in Minato-ku. The place was elegant and appealed to well-heeled Western visitors. Stunningly beautiful. It was recommended to me by a William & Mary history professor, A. Z. Freeman, who had Shizuko Skoglund translate correspondence for him. He was a collector of samurai swords and was a frequent visitor to Japan. He suggested the International House, believing it would help me get my feet on the ground. Useful advice.

Upon my arrival by taxi from the Tokyo terminal, an eager bellhop carried my duffle bags to my room, then swept open the sliding panels in a grand gesture, so I could behold the expansive gardens just below. "*Kirai desu*," I said. The bellhop looked puzzled by my comment, so I said it again to be sure he heard me correctly, "*Kirai desu*!" Following my emphatic repetition, he bowed solemnly and backed out of the room, leaving me to gaze at the gardens. Yes, they were beautiful. That's what I was trying to tell him. Hmm, let's see, how should I have said that? Oh yes, the right word is *kirei*, meaning "pretty" or "nice." Close enough to mean "beautiful"? But I hadn't said *kirei desu*. Instead, I used *kirai desu*. The words look nearly identical, but *kirei* is pronounced ki-*ray* (rhyming with day) and *kirai* is pronounced ki-*rye* (rhyming with eye). Ray? Rye? Had I mixed them up? Yes, I surely had. That slight slip in pronunciation changed the meaning of what I had hoped to say into something altogether different. *Kirei desu* does mean "it's beautiful" or "pretty" or "nice" or "tidy" (all complimentary), whereas *kirai desu* means "it is hateful." So, when the bellhop was showing me the garden view from my room, I told him I disliked it—hated it—and to be sure he understood, I repeated it until he ran away. I was off to a good start my first day in Japan. *Oy vey* (which, in Yiddish, means "I am not a polyglot!").

As I sat in my room, wondering if I should chase after the bellhop to explain my mistake, I recalled my reaction to something A. Z. Freeman had told me just before I left Williamsburg for Japan: "Remember, they despise us." I was disturbed by his comment and thought it uncharacteristic of this affable gentleman. Well, maybe they despised *him*, I supposed, because *he* was hauling away their national treasures, whereas I'm just going there to study moths. I didn't say anything like this to A. Z., of course. We were from different generations. He specialized in military history, and World War II was a recent memory. As for me, I had been a baby during those horrendous years, and now I was a field biologist who lived in the woods. Why should they despise *me*? Perhaps my saying "*kirai desu*" to the bellhop welcoming me to his beautiful hotel was a start. I vowed to be more careful in the future. Mind the jet lag!

That Monday afternoon, on my first day at the International House in Minato-ku, I wasn't in my room for long when the sliding panel doors began to shake. Already? I had asked Hiro what the chances were that I might experience an earthquake, or at least some tremors, while I was in Japan. He answered by asking what I might say to someone planning to visit Virginia for four months about the chances that it might rain while they were there. Well, it hadn't rained yet in my first few hours in the Far East, but already the ground shivered enough to shake the doors in my room. It didn't last long, but I figured I'd call home to tell Cathy about my safe arrival, my horrific *Nihongo*, and the earth tremors. When she answered the phone, she sounded a bit groggy and confused, and then asked me what time it was.

The next morning, after an American-style breakfast of fried eggs and coffee at the International House, I attempted to ring Professor Kitagawa to set up a time for our first meeting. I had practiced how to introduce myself by phone and how to ask to speak with him. I had the university's number and could dial 717-0111, but I also knew I'd be asked where to direct my call. Kitagawa's extension was 3521, so I practiced saying those numbers: "*san go ni ichi*." But if he wasn't in, then I'd ask for another departmental faculty member, Yoshiaki Fuyama, at extension 3543, by saying "*san go yon san*." I also practiced "*sei butsu gaku no Kitagawa-sensei, kyo oshitsu o onegai shimasu*," or words to that effect.

Despite all that preparation, I was really hoping Kitagawa might answer the phone himself, because I suspected he spoke English rather well. He had been a postdoc of the greatest *Drosophila* evolutionist of them all, Theodosius Dobzhansky, at Columbia University in New York City. Although Dobzhansky had a slight Russian accent, his English was impeccable, and his every utterance came out in publishable form. Unfortunately for me, on May 31, 1988, Professor Osamu Kitagawa didn't answer his phone. Instead, I was greeted by a young woman's voice in *Nihongo*. "*Moshi, moshi*," she said sweetly. Although I had rehearsed for this very conversation, my mind raced back to the bellhop at International House, and I worried that I might embarrass myself yet again. So I asked, "*Eigo hanashimasuka*?" I'm not sure if she answered my question, but a flurry of words followed, to which I responded with "*wakarimasen*." We had quite a long conversation, repeating ourselves several times, but finally I hung up and wondered what to do next. After about a half hour I rang again, hoping Kitagawa might answer this time. Nope, it was the same woman, and we had the same pleasant conversation. Dang, and after all of my rehearsals on just what I'd say. Finally I decided to go to the hotel desk and ask the receptionist if she'd place the call for me. As her job required her to deal with foreign visitors every single day, she sounded quite fluent in English. She learned from her call to *Toritsudaigaku*'s Genetics Department, on my behalf, that Kitagawa was not yet in the office, but he would like to invite me to lunch. We'd meet with the dean that afternoon, and then with the university's president. Meet with the president? Well, that would be a new experience for me. Better wear a tie.

I thanked the receptionist for relaying that information to me. While I still had her undivided attention, I took the opportunity to explain to her—using English, not Japanese—my earlier miscommunication with the bellhop when I had used *kirai* instead of *kirei* in attempting to express my appreciation for the view of the garden from my room. I was asking her to please relay my sincere apology to him for my faux pas. As I spoke, her amiable smile was replaced by a look of concern, and when I finished she said politely in perfect English: "I am so sorry for this unfortunate incident, sir. I will see if we can change your room to one that pleases you."

"No, no. *Iie*!" I said and tried once again to assure her that I quite loved the garden and my room. As I went on and on in this vein, I overheard another Japanese woman speaking in German to a group of Germans gathered just down the hall. Of course, there were people at this hotel from all over the world—from France, Italy, Spain, you name it. All sorts of languages could be heard at the International House of Japan. It was a regular Tower of Babel.

28 | Reception

Although the train station was only a few blocks from Tokyo Metropolitan University (TMU) in Setagaya-ku, not far as the crow flies from Minato-ku, there was no direct connection. To go there by train (or subway), I'd have to travel to Shibuya and change trains. I wasn't yet prepared to do that, although I would soon learn, with experience and much local input, to move through Tokyo by train like a native—lost only about half the time, also like a native. On my first day officially on the job, however, I didn't want to risk getting lost and being late.

When I arrived at the campus, my heart sank at its grim appearance, looking more like a prison than a seat of higher learning. Most of the buildings were gritty gray and appeared to be constructed from poured concrete. The grounds were unkempt, and the lawns overgrown. I was greeted by a uniformed security guard at the entrance to the biology building and directed by him to the third floor, where I'd find Kitagawa's office. I was expected.

The hallways were crowded with equipment, some old, some new, some in use, some waiting (probably for an appreciable time) to be hauled away. One had to zigzag to walk down the halls. The labs inside were even more jammed, with equipment scattered everywhere: refrigerators, incubators, centrifuges, computers, tables holding scales or microscopes or electrophoresis gear. Crammed in between, wherever space allowed, were desks for graduate students.

Kitagawa, as a chaired professor and active head of the lab, had an office, but he shared it with the former head of the lab, who, in his retirement years, continued his research on flies. Kitagawa introduced us with a cursory ceremony, and then the emeritus professor returned to his work with great urgency. His remaining time as an active scientist was running out. I'd see him occasionally over the next few months as he scurried through the halls, always wearing a lab coat and a necktie, ferrying fly vials between his desk and incubators in the hallway.

As soon as we bowed in greeting (I don't recall if we shook hands), Kitagawa apologized for not having a spot for me in his office. That had been his intention, but it would mean finding another place for the emeritus professor to do his work. No need to apologize, I assured him. (Frankly, I was relieved not to have the desk right next to his. I craved more privacy.) Kitagawa then took me into the lab to show me where I would be located. As we entered, the graduate students, sitting here and there at their desks, rose and bowed to us. Wow! This is great, I thought. Such respect—*reverence*—I never get at home! The work site that was to be mine was just like the rest, surrounded by equipment and other desks. I was now a member of a *Drosophila* genetics lab once again.

There were flies everywhere. That struck me as odd. One of my former graduate school professors was Terumi Mukai, who had returned to Japan after spending a few uncomfortable years in the Genetics Department at North Carolina State University (NC State). He had inherited his lab technicians from his predecessor, Ken-ichi Kojima, who had left for the University of Texas as the rising star in experimentally demonstrated frequency-dependent selection (FDS). Mukai's constant complaint was that his technicians allowed too many flies to escape, and he worried that visitors to his lab, seeing so many loose insects, wouldn't trust the data from his meticulous, painstaking experiments assessing the cumulative effects of polygenic mutations on fitness—differences of decimal places. Mukai ultimately did nearly all of the critical lab-bench work himself, because he didn't trust his careless technicians. They, who had loved assisting the glamorous Kojima, hated working for the tedious Mukai. So I was flabbergasted by all of the loose flies in Kitagawa's lab. No one seemed to notice them but me. I had never seen anything like it, except maybe in the even more glamorous Dick Lewontin's lab at Harvard.

Kitagawa and I went to lunch on my first full day in Japan, and there we met Yoshiaki Fuyama, another *Drosophila* geneticist who was then an associate professor in Kitagawa's department. I should mention that I never called any of these people by their given (or what we call "first") names. That just is not done in Japanese culture. First names are strictly for family members or for very close personal friends who grew up together. Fuyama remained Fuyama-san, and the same for Kitagawa-san (or, in his case, as head of the group, I called him *sensei*, much to his amusement). Even Hiro was referred to as Asami-san—or, occasionally, as Asami-kun, a term of endearment—by his fellow students still there. I became "Gu-ran-to-san," never to have "Grant" pronounced as a single syllable when it could easily be made into three. Sometimes people mistakenly called me "Ba-ru-su-san" (for Bruce-san), because my given name appeared first on my *meishi* (business cards), so it was assumed to be my surname. Because of the potential for confusion caused by the Western custom of placing family names last and the Asian practice of placing them first, many Japanese capitalize all the letters in their family name. For example, Kitagawa's name would appear as Osamu KITAGAWA and mine as Bruce GRANT. In any case, A. Z. FREEMAN advised me to have business cards made up for my visit to Japan, because everyone exchanged them, including professors—a practice unheard of among American scientists, at least those I know. Mrs. Skoglund told me that if I didn't have a business card to exchange with others, "they'll think you are a nobody," emphasizing that status is very important in Japan. So she made up my cards, hundreds of them—English on one side, and Japanese on the other. I handed out four of them on my first day at TMU. Except for those, the rest are still in the box.

While at lunch, we chatted politely about this and that. Kitagawa, tall and lean, neatly attired in a suit and tie, was much older, more formal, and soft-spoken. Fuyama, somewhat rounder and shorter, with an open collar and sandals, was delightfully irreverent about nearly everything. We got along famously. I liked both men very much and felt gratefully relieved that Hiro had arranged for them to accept me as a member of their lab. Indeed, without them I'd have been lost, both figuratively and literally.

After lunch we went back to Kitagawa's office to pick up some things he wanted to take to our upcoming meetings with the dean and then the president. One of them was a copy of the manuscript of the paper Rory Howlett and I had written, which I had sent to Kitagawa to let him know what I was up to. By then our article was in press, and he probably had a copy of the galley proofs. The other thing he wanted to show to the dean and the president was the newly released second edition of *Population Genetics and Evolution*, a book by my former mentor (L. E. "Gene" Mettler), with coauthors Tom Gregg and Henry Schaffer added. Mettler had asked me to revise his chapter on natural selection, to bring it up to date—in particular, his treatment of industrial melanism—so of course I included a hefty dose of my own work and the recent declines in melanism documented so carefully by Cyril Clarke.

Seeing the book here in Japan pleased me and reminded me of a pleasant excursion to Raleigh I had taken the year before it was published. The official reason for that trip was to present a seminar about peppered moths to the Entomology Department at NC State, but I took the opportunity to drop by the Genetics Department to visit Mettler. He had a grand time teasing me about how his revised chapter on natural selection now mentioned Bruce Grant more than it did Charles Darwin. Surely that's not true! Either way, Kitagawa wanted the dean and the president to understand, by showing them my upcoming paper with Howlett and the recent textbook treatment of the research I was doing, that I was not a tourist visiting Japan, but had come to extend this important, classical work.

It was a short walk to the main administration building, where a secretary directed us into the dean's office. After formal introductions and bows, I was invited to sit down. All of the furniture was draped in white linens, as if in a warehouse. The secretary wheeled in tea and patisseries on a cart. We chatted about my trip and my initial impressions of Japan. The dean then apologized about the physical appearance of the university, explaining that they were preparing to move to a new campus, which was in its final phases of construction. The real estate in Setagaya-ku, where TMU had been built, had greatly increased in value and would be sold to developers. The entire university would be relocated well away from central Tokyo, to a site where land was much cheaper. As all of the buildings on the present campus would soon be razed, little effort was put into upkeep. Ergo, the dilapidated condition. That prompted me to ask about some active construction work I had noticed on entering the biology building. They were renovating the women's lavatory to include a rest lounge. This seemed an odd thing to do at a time when they were planning on demolishing the building within the next year, but apparently this particular upgrade was mandated by law, with no option. Neither the dean nor Kitagawa revealed any hint of disapprobation. The law is the law. Period.

Everything was lovely, and Kitagawa and the dean seemed perfectly relaxed. I, however, wondered what our meeting was all about. When were we ever going to discuss business—such as when and how might I get somewhere to trap peppered moths? All in good time, my boy. They'd get around to that. It would be unseemly to jump right into things. This meeting was just so we could get acquainted. Before leaving, Kitagawa did present the dean with a copy of my manuscript and showed him the photos of moths, wearing paper collars, that had been published in Mettler's book. The dean smiled approvingly and welcomed me to the university.

The meeting with the president went similarly, except his office was much bigger and had more furniture, which was also draped in white linens. More tea and patisseries were wheeled into the room, this time by two secretaries. I noticed that each time the secretaries left, they backed out of the room, bowing the whole time. I was getting the royal treatment. Indeed, the president explained to me that I, as a visiting professor, had all the rights and privileges of full-time faculty, except for the right to vote. I was entitled to use the full resources of the university, as might be appropriate and available to me—except, of course, for my salary and expenses, which were not included. Fair enough. I knew that. Because of my TMU affiliation, I would be permitted to trap moths at various field stations throughout the country without being charged lab fees. I would simply pay for my accommodations. This was a wonderfully generous arrangement. I could not have hoped for more.

But more was in store. The department had arranged a welcome party in my honor, to be held on the expansive flat roof of the biology building, following the regular workday on Saturday. Yes, they all worked full days every Saturday, but most people took Sundays off. This reception was scheduled for a few days after my arrival, perhaps to allow me to get my feet on the ground, but more likely because the next day people could sleep in. I had no idea that it would be such a gala event, and that so many people would attend—individuals I had not even met yet. Well, I suppose that was the point. We'd all get to know each other then. For the few days preceding the party, Kitagawa kept reminding me of it and asked me to be sure to stay over on Saturday afternoon for the occasion, as it was being held in my honor. Without my presence, the event would be pointless, or so he insisted. How could I refuse such an invitation? Besides, where else did I have to go?

So, up to the rooftop I went after work on Saturday. The weather could not have been better, and we had an incredible view of the campus and the city beyond. There were trays and trays of food of all sorts. Japanese food not only has to taste good, it has to look spectacular. Tremendous attention is paid to presentation, with various items placed on serving platters in precise geometric patterns. Absolutely kaleidoscopic! I would hate to be the first one to dive into any such display, for fear of destroying its appearance. That turned out to not be a problem for me, because there was a throng of hungry graduate students on hand who had no such qualms. They dived right in. *Itadakimasu.*

I selected various items that looked familiar to me from my experience in Virginia with Mrs. Skoglund, but here I was being watched. The person observing me, out of the corner of her eye, was Yoshiko Tobari. We had actually met a number of years ago, when I was a graduate student and she was a postdoc of Ken-ichi Kojima's at NC State, but I couldn't claim that we knew each other. Kojima was a world-renowned figure in population genetics, and his lab was full of postdocs, especially from Japan. Ultimately, Tobari returned to her native Japan and, through her own hard work, established an international reputation. It was rare at that time, and even now, for women scientists to rise to the top ranks of academia in Japan. Since Tobari was a full professor at TMU, it was abundantly clear that she was highly respected at home and abroad. She wasn't watching me because she remembered me from NC State, however. She did so because she was the hostess for this welcoming party. She worried that I was missing out on the best food Japan had to offer. She noticed that I was choosing cooked tidbits and skipping the sushi and sashimi. How observant she was! I deliberately avoided any and all uncooked fish. Not for me, thank you! Of course, I knew long before coming to Japan that *Nihonjin* thrive on raw fish. Those who can afford it eat it daily. But as I had never before eaten this popular delicacy, I was afraid to try it. I was worried that if I did attempt to swallow the uncooked stuff, I'd gag on it and maybe even throw up. That'd be a bad way for me to behave at a party given in my honor. While I didn't think my status was sufficient to result in an international incident, I did realize that I represented my country and was expected to show respect for the culture of my hosts. I could no longer just pick and chose what I thought were safe items of food to ingest when Professor Tobari said, "But, Guranto-san, you haven't tried the sushi."

"Er, well, that's okay, I'll have some later," I protested lamely.

"No, no, have some now," she said. "You'll love it. Here, try some of this," she said as she indicated a particular platter. "It's easy for first-time experience."

Tobari completely understood my predicament, and she knew exactly which sample I should try first, to get me over the hump—or avoid causing me to gag.

I said to myself, "Well, here goes, I just hope I don't retch in front of all these smiling people." Then I reached down and picked up a piece and popped it into my mouth. I swallowed it as quickly as I could, hoping to not taste it. But, try as I might, I did taste it. Suddenly I found myself saying, "*Oishii desu*" ("It's delicious")! I repeated this, and I meant it. I was not just saying *oishii desu* to be polite. I flat out loved it. So I tried another piece, then another, and went on to try some other samples, until I had gobbled up a small fortune in raw fish. That welcome party turned me into a sushi hound, and I still am. To this day, I am grateful to Tobari-san. *Gochisosama deshita.*

At this point, the party was only beginning. There was also a lot of *biru*. That I knew how to drink, and my hosts saw to it that I had great quantities. I would barely take a sip out of my glass when someone would immediately rush over to top it up. Japanese beer is really good. I know their traditional drink is sake, and there might have been some at this party, but I never developed a taste for it. I certainly didn't miss it, as the supply of *biru* seemed endless. I was flattered by all of the attention and happy that everyone seemed to be having a good time.

Very early on, I noticed three young women standing nearby who were giggling practically nonstop. I was curious and asked Fuyama what was so funny. He said they were drunk. "Drunk?" I asked. "How could they be drunk so soon?"

"Oh," said Fuyama, "they do not have enzymes."

Well, like all of us, of course they have enzymes. What Fuyama was referring to was a particular form of enzyme called alcohol dehydrogenase, or ADH. As we were both geneticists, he was aware that I knew there is considerable geographic variation in the frequency of alleles of ADH genes in human populations, with rather striking differences between Asians and Europeans. Simply put, some people get drunk very quickly, while others tolerate much larger amounts of alcohol. Fuyama, in his entertaining way, was saying these young women lacked the "right" form of the enzyme and therefore had become inebriated very quickly. "Fortunately," said Fuyama as he refilled our beer glasses, "we have the right enzyme."

"*Kanpai*," I toasted, but silently wondered who was more fortunate—those who can tolerate large quantities of alcohol, or those who get drunk on a single beer? I'd save a lot of money if I had different enzymes.

Still more welcoming events followed the rooftop party, although they were spaced out by a few days of lab time. The next gala affair was set at Kitagawa's home—an evening dinner party, for which I was to be the guest of honor. To get me there comfortably, Kitagawa assigned one of his graduate students, a young man from Burma named Soe Wynn, to serve as my guide. (This was 1988, and the official name change of Burma to Myanmar was still a year in the future.) Soe Wynn was selected to escort me because his English was especially good and, as he had been at TMU for several years, knew his way around Tokyo. So off to Kitagawa's house we went. Along the way, Soe Wynn coached me about how the dinner party would unfold: "The drinking will start almost immediately upon your arrival."

After warm greetings of welcome by Kitagawa and his wife, I presented them with the bottles of Johnnie Walker Black Label, to which Kitagawa smiled broadly and said, "My favorite! How did you know?" I was then introduced to other members of the Kitagawa household, as three generations all lived in the same quarters, which was the norm in Japan's teeming cities. The only other guest was Fuyama. After several rounds of drinks and toasts, Mrs. Kitagawa began serving the meal. I cannot recall how many courses there were, but each was beautifully presented and included more meat than I had ever seen at any single meal. Because they knew I was an American, they had assumed I loved to eat beef, so there were lavish stacks of it, prepared in a variety of ways. I know too little about fine cuisine to describe the imaginative courses, but the cost alone of so much beef in Japan suggested that they had pulled out all the stops to make me feel welcome.

The next morning, back in the lab, one of the graduate students asked me if I had had a good time at Kitagawa's dinner party. Absolutely splendid, I assured him, but I made the mistake of adding that I had left my umbrella in the taxicab on the way home afterward. It was raining that morning, and while an umbrella would have come in handy, what really bothered me was that this umbrella had been a gift from Elspeth, my younger daughter. She said that it rained a lot in Japan, so I'd need it.

I didn't think much about my comment for the rest of the day as I worked at my desk. Then Kitagawa came up to me and asked, "Is this your umbrella?" I accepted it from his outstretched hand and said it most certainly was. I recognized the brand name on it, Gitano. He then added that the name meant "gypsy" and suited me, as a traveler in a foreign land. But how did Kitagawa recover my umbrella? I was positive I had it when I got into the taxicab the night before. Indeed, I had. The graduate student to whom I had mentioned this earlier in the day reported my carelessness to Kitagawa, who in turn called his wife, who contacted the taxi company, retrieved it from their lost-and-found department, and then brought it to the university. Who does that? My Japanese hosts did.

29 | Around Town

The various welcoming parties and introductions were interspersed with several days spent in the lab and guided tours of places in Tokyo where I might purchase materials I would need to construct Robinson-style moth traps. Fuyama took me to a gigantic department store in Shibuya, the busiest shopping district in Tokyo. In the process, I learned how to use the trains to get around the city. Tokyo was then the most densely crowded place I'd ever been to, until later visiting cities in China and India. Shibuya has them all beat in terms of being upscale and bristling with excitement. As we crossed busy streets, surrounded by people on all sides, Fuyama commented that more people always seemed to be going in the opposite direction from him. He wondered where they were heading. Surely they must know something he didn't, and he considered following them to find out. This was his usual banter, as I came to discover. Observational humor! What I noticed was that no matter how many people were coming toward us, and how many were moving with us, the two immense masses passed by each other without anyone bumping into anyone else. They were masters at moving smoothly through a throng. Later I mentioned this to an Australian cousin, who laughed, saying, "An Australian can't walk through an airport without bumping into every single person he passes in either direction." Is this true?

But there are places in Japan where close personal contact cannot be avoided: riding subway trains in Tokyo during rush hour. There are actually attendants on the platforms—wearing white gloves, of course—whose task is to physically push people into the train cars before the doors close. I made the mistake of taking the subway early one morning while carrying my two duffle bags. The platform attendants not only squeezed me and my bags into a packed car, but no one grumbled, complained, or uttered a whimper. That's simply how life is during rush hour. Otherwise, physical contact is assiduously avoided in public.

Inside the multistoried department store, we were bid "*Yokoso*" by smiling young women wearing snappy uniforms, and later told "Thank you," with invitations to return soon. Their greetings were practically sung, in sweet, cheerful voices. They made me want to buy something, anything, just so they'd be rewarded for being so nice. In Japan the consumer is king and is made to feel that way. Even at petrol stations, which I frequented when using generators to run moth traps away from sources of electrical power, at least four uniformed attendants would come running out to clean a car's windshield, check the engine oil and air pressure in the tires, and pump the fuel. And, as the customer drove away, they'd bow deeply in unison, like stage performers, and call after him, "*Itterasshai*."

To my ear, it seemed as though men and women spoke different languages. It was all in Japanese, of course, but masculine and feminine linguistic styles were conspicuously different. For starters, the different genders have certain vocabulary terms reserved just for them, such as in ways to apologize. One way to say "I'm sorry" or "excuse me" is *sumimasen* and another is *gomenasai*, among others. But men don't use *gomenasai*, because it's regarded as being too feminine. I don't really know these subtleties, and I employed those apologies interchangeably without batting an eyelash. Instead, I was irresistibly drawn to the lyrical intonation of women's speech, compared with the rapid-fire staccato of men's. Even when males are not angry or impatient, they often sound that way. Human language! How intricately diverse. The people I envy most are those who are fluent speakers of more than one language. I can ask many questions in several languages, but I rarely understand the answers.

As I learned the ropes about how to travel by subway train and bus, I made several excursions on my own to shop for various bits I thought I might use to put together moth traps. Some of my trips were just for fun, to get to know the city—such as visiting Tokyo's version of the Eiffel Tower, or seeing the highrise buildings in Shinjuku, built on special tremor-resistant foundations—but mostly I stayed at the lab and constructed moth traps that I tried out on campus at night. Despite the close proximity of competing lights, a variety of moth species were coming to my traps, assuring me that my equipment worked. Alas, not a *Biston* among them.

I also moved to more affordable accommodations, a hotel called the New

Hideoka in Sangengaia. I had to take the subway to and from work each day, and after initially missing my stop on the way home, I learned to listen carefully to the conductor's voice announcing the stops. The one just before mine was Yutenji, so when I heard the announcement, "Yutenji-desu, Yutenji-gozaiamasu!" I knew the next stop, Sangengaia, was mine. When people asked me where I was staying, I sometimes told them Yutenji, but I pronounced in such a way that they didn't understand where it was. I said "Yuuu-tenji," emphasizing the first syllable, as I had heard the train conductor pronounce it. Years later, when riding on Amtrak in Virginia, it dawned on me why people in Japan looked at me strangely when I told them where my hotel was. The conductor announced the Richmond stop as, "Err-richmond, Err-richmond, Virginia, is the next stop. Err-richmond!" So if someone asked me where I lived in Virginia and I answered "Err-richmond," they'd probably look at me as quizzically as the Japanese did when I said "Yuuu-tenji." The morale of this story is not to learn the pronunciation of places from train conductors.

My room in Sangengaia's New Hideoka hotel had a Western-style toilet, complete with illustrated instructions for its use, and a TV. Fortunately, I knew how to use both. I found an educational channel that taught English to native Japanese speakers. I watched it every time I was in Tokyo. One thing I noticed was that on certain days they had American instructors, and on other days they had British speakers. I wondered if this was to help the students learn to distinguish between them. As Japan does an enormous amount of business with both the United Kingdom and the United States, knowing both styles of English might be useful. I asked my English-speaking Japanese colleagues in the fly lab if they could recognize whether a speaker was from Britain or America by his or her accent. Not a single one of them could. At least I think that's what they said. I was not surprised. I can't tell the difference between the German spoken by an Austrian, a Prussian, or a Bavarian, although German is one of the two foreign languages I was required to learn for my PhD. The other was French. I can, however, tell the difference between German and French.

My Japanese colleagues encouraged me to attend a seminar one afternoon, where the talk would be given in English. They were hoping I might be able to fill them in on the bits they'd miss. The visitor was from India, and sure enough, as he was from the most populous English-speaking country on the planet, he gave his talk in that language. I sat as politely dumbfounded as my Japanese colleagues in the audience. Afterward, back at the lab, people asked me what he said. I replied, "I dunno." "But wasn't his talk in English?" I was asked. "I'm sure it was," I said. "But his accent was so thick—unfamiliar to me—that I could not understand what he was saying. And, what made it worse, he spoke so rapidly that I couldn't follow him." My Japanese hosts were very relieved. They had worried that their inability to understand the speaker's English was their fault. It made them feel a lot better to learn that an American also couldn't understand him. Happy to be of service!

As the days passed, my anxiety grew. I was losing precious time that I needed to spend in the field. I did not know just when or where peppered moths might fly in Japan. So far I had not caught any on the university campus in Tokyo, albeit the overall numbers were understandably small. I was eager to get well away from city lights to make *big* hauls. To make matters worse, I was hearing about people in other parts of Japan who were catching *Biston* while I was languishing in Tokyo hotels. The source of this distressing news came from a botanist at TMU, Dr. Masamitsu Wada, who was in contact with Dr. Rikio Sato, who was running moth traps elsewhere in Niigata. "Has he caught any melanics?" I eagerly asked Wada-san. Apparently not, but at least he was catching Japanese peppered moths in goodly numbers, while I was not. Wada-san was a friend of Hiro Asami and dropped by from time to time to chat. He was sympathetic with my plight and agreed that it was time for me to get out of Tokyo.

Kitagawa certainly sensed my growing impatience and informed me that soon we would get started. He had arranged for me to meet Professor Hiroshi Inoue, the "dean" among Japanese lepidopterists, who had authored the book *Moths of Japan*. We would travel by train the next morning to Otsuma Women's University to meet with Dr. Inoue in his office. If anyone in the country could set me on the right course, it would be Inoue!

We changed trains several times on our long journey, by which time I was hopelessly lost. At one of the busier platforms between trains, Kitagawa stopped by a shop to purchase some patisseries as a gift for Inoue. I offered to pay for them, as this visit was for my benefit. "No," Kitagawa insisted. "This is *our* custom. Not *yours*." I dared not argue with him. Kitagawa was *sensei*. Not arguing with *sensei* is another Japanese custom.

During our train journey, Kitagawa and I got to know each other better. His formality loosened, and we chatted about acquaintances we had in common. One such person was my old professor from NC State, who had joined the faculty at Kyushu University. Kitagawa had once held Terumi Mukai in high esteem as a colleague but apparently became disappointed that Mukai had assumed the role of "guardman," as Kitagawa put it sourly, for Motoo Kimura's *neutralist* theory. Kimura, the shogun of this theory, was securely ensconced at Japan's National Institute of Genetics and was brutally critical of his opponents. Kitagawa, as a former disciple of Dobzhansky, was a confirmed *selectionist*. While the scientific merits of these titanic arguments have been thoroughly reviewed elsewhere by legitimate experts and are of little relevance to the evolution of melanism in peppered moths, I was struck by the fervor that existed between the opponents. Japanese selectionists were alive and well and were not kindly disposed to Kimura's bullying.

When we arrived at Inoue's office, I was impressed by the neatness of the place and its simple elegance. No white linens draped over the furniture! That Inoue had such a large, unshared office at a small college suggested that he was valued there. He seemed to have his very own secretary, also unusual for faculty not holding an upper administrative rank. Within minutes of our introductions and Kitagawa's presentation of the patisseries he'd brought, in came the secretary with her tea cart. As I had come to expect, a lot of time was devoted to polite chitchat. First, I conversed for a few excruciating seconds in *Nihongo*. I was pretty good with greetings but soon ran out of ways to ask "How's it going?" Inoue expressed what seemed to be genuine delight that I could converse at least *sukoshi* in his language. Fortunately for me, he could speak fluently in mine.

Once we finally got down to business, Inoue told me that he had never heard of anyone ever collecting a melanic specimen of *parva*, the Japanese subspecies of *Biston betularia*.

30 | In the Field

Finally I was poised to collect moths in Japan. To say I owe this to Kitagawa and Fuyama would be an understatement. Negotiating the convoluted Japanese bureaucracy was well beyond my comprehension. I would still be in the New Hideoka hotel in Sangengaia, watching language lessons on TV, had it not been for the two of them.

What they did was make some phone calls. These were calls I could not have made even if Mrs. Skoglund were sitting by my side. I was stumped about what to do and how to proceed. Without Kitagawa and Fuyama, I might as well have headed back to the airport and gone home. But no. They were calm. They knew how things worked. They had patience. They were Japanese and understood the intricacies of their own culture. I didn't. I was a bull in a china shop. I'd have gotten absolutely nowhere on my own, even though I speak *Nihongo*—well, *sukoshi*.

We were in the car heading to Takao, a subdivision of Tokyo. Kitagawa was driving. Fuyama offered me the front passenger seat but I refused the honor, insisting that he and Kitagawa could converse more easily. I listened from the back seat, picking up the odd word, probably more than they realized but not as much as I pretended. After about an hour or so on heavily congested roadways coming out of the Setagaya section of Tokyo, we arrived at a secluded national forest preserve with strictly enforced access. This was a woodlands research facility, closed to the public. Might I be allowed to run moth traps here? On my own, begging, the answer would be "No, *iie*," delivered politely among a fusillade of synonyms, including "impossible." But Kitagawa had influence and diplomatic skills.

By the time we arrived, it was late afternoon. Most of the regular staff were on their way home, but we met with the chief administrator in his office. As I had come to expect by now, the furniture was covered in white linens, and secretaries plied us with patisseries. No direct "Let's do this" came from Kitagawa. Fuyama, knowing his own rank, was mostly quiet. I, for lack of sufficient *Nihongo*, was speechless. Ultimately, after much amiable jabber, we took a walk down to the rear parking lot. The talk went on. I understood practically none of it. We then got into Kitagawa's car and drove off. What had happened?

Kitagawa waxed philosophical, saying, "On the phone they wouldn't give us an answer, but upon seeing our faces they agreed to let you trap moths here in the national forest."

"Wow!" said I. "When?"

"Tonight," said Kitagawa.

I was truly dumbfounded by his answer. I had been in Japan long enough to know that negotiations take time, patience, and guile. I had not been in Japan long enough to realize that decisions can also be made on the spot when the negotiators are skillful. I learned more about Kitagawa that day than about stereotyped Japanese traditions.

But there were "under the table" provisos. Until the official paperwork could be processed (taking several days), my trapping efforts would have to be kept quiet—unofficial. This meant setting up and unpacking my moth traps "out of sight" during regular operating hours. Nine to five? Whatever. I was happy with this arrangement and eager to get started. Tonight!

Fortunately, I came fully prepared, with my moth trapping gear stowed in the trunk of Kitagawa's car, including a Honda portable electric generator that ran on gasoline, kindly lent to me by Dr. Ono, a botanist at TMU. While I had not been so optimistic as to think I'd start trapping moths that very night, I had intended to move to Takao in the hope that we'd work out something. With Kitagawa's and Fuyama's help, we spread out three moth traps in the mountain forest, well away from city lights. The traps were connected by long extension cords to Dr. Ono's electric generator. After getting the generator fired up and checking that all of the mercury vapor lamps were shining brightly at the mouths of the traps, Kitagawa and Fuyama then took me to a hotel in Takoa, where they had arranged for my extended stay. With gleeful smiles and figurative pats on the backs (no touching), my colleagues then returned to their homes in Tokyo. It was summer, and still broad daylight. I felt terrific—finally getting to work and grateful to my hosts for making this happen. *Omedeto gozaimasu*!

The next morning I was up at the crack of dawn, like any self-respecting moth trapper. But the national forest would not open its gates for a few more hours. So I drank tea, ate some rice, and paced around until I was allowed in. Smiles, bows, and "*Ohayo gozaimasu*" all around. Then I hit the road on foot, up into the woods where my moth traps were waiting for me.

It was a steep climb along a dirt road, but I easily took it in stride. I was in shape then, and eager. As I approached the section where I thought I'd set up the traps, my heart started racing with excitement. Then I grew dimly aware that something was missing. Sound! What I was not hearing on my approach to the trapping site was the generator's motor whirring in the woods. This would be a familiar sound I expected to hear from a distance, but as I grew closer, there was nothing but silence! Is it possible to say the silence grew louder? Then I spotted my first trap, sitting exactly where I'd put it—in a small clearing between some trees. Its light was off. The lights on all three of the traps were off! The generator had run out of gasoline and stopped operating long before I got there. There were moths in all three traps, which was clear evidence that the generator ran for at least a short time after nightfall, but the catch was small. It was obvious that the lights in the traps were not on for most of the night. By starting the generator before the national forest closed at 5 PM meant that by the time it got dark, this little generator had run out of gas. The Honda generator I used back home was a huge model, compared with this one, and would easily run all night—and then some. But the portable model I'd borrowed from Dr. Ono had a much smaller tank and was designed to operate for only a few hours before it needed to be refueled.

What was I to do? I had to start the generator at 5 PM, four hours before it got dark. With that arrangement, I'd only be getting in about an hour of trapping before it shut down. I'd either have to buy myself a new, bigger, longer-running generator (which, besides the expense, would seriously impede its portability on public transport) or come up with another plan.

After smiles, bows, and "*De wa mata ashita*" all around, I'd leave the national forest at closing time, after setting up my traps—though *not* starting the generator. That would happen later, *after* dark, when the regular staff had all gone home, leaving only security guards at the entrance.

Clearly I would not be allowed to access the national forest after they locked the gates to the public at closing time. For me to go in after dark meant I'd have to break the rules of our agreement and find a way to sneak in. What I decided to do, in the name of science, was no doubt illegal, and for that I apologize to my gracious hosts, who trusted me. But I was desperate and decided to climb over the security walls enclosing the facility to start my moth traps under cover of darkness. Even then, the generator didn't have the capacity to run all night, so I'd return, going over the walls yet again, to refuel the generator halfway through the night. I worried how I might explain my activities to the security guards—or police—had they caught me. Would they see me as a courageous ninja moth man and let me off with a warning, or would they put me in jail? I alternately pictured myself as the stealthy cat burglar character in the *Pink Panther* and the bumbling Inspector Clouseau.

Once the generator was running all night, the moth catches at Takao were robust. I had huge numbers, representing many species. As I am not a lepidopterist, I was not able to compile a species list for these catches. Yet for some reason, Kitagawa thought I should be doing exactly that. For my purposes, it would be a complete waste of time. Sure, at a glance I could identify to the family level many of the larger, common moths that came to my traps. That was pretty easy to do. Then again, there were lots of LBTs (little brown things), about which I knew nothing. Such taxonomic assignments were not remotely related to what I was interested in studying. I reminded Kitagawa that I was neither a lepidopterist nor a taxonomist. I was a population geneticist, looking for the parallel evolution of melanism in peppered moths. While one cannot help being drawn in by the amazing diversity and beauty of moths that exist in our world and show up in trap samples, each species having its own fantastic evolutionary history, my mission was much more narrowly focused: on just one thing and only one species. That was all my personal expertise would allow. Science is a huge enterprise, but we practitioners must restrict our attention to certain bits of it to be effective. Still, wherever I trap moths, people invariably draw my attention to some big, spectacular moth they might have seen the previous night, often saving the specimen for me. They assume, because I am hunting for a certain moth, that I am a moth collector. It's easier for me to say "Great! Thanks!" when they show me some moth than to explain why that moth is not the one I'm seeking. Everywhere I go, people try hard to be helpful, because I haven't taken the time to explain my special interests. That's quite okay. I did not expect a "there are lots of moths here" from Kitagawa. He knew exactly what I had come to study, so I was puzzled when he suggested I keep a log of what species I caught each day in my traps. What? Why should I do that?

To this day, I cannot say whether Kitagawa, in his negotiations with the officials at the national forest, might have offered them my services to provide a moth survey of their woodlands. Some species are serious pests to the trees they had there, and such information from a moth survey would be useful. Perhaps, I wondered, was that why they so quickly granted me access to their forest? But I was the wrong guy to conduct this sort of work. While they didn't know that, surely Kitagawa did.

Kitagawa and I both shared a background working with *Drosophila*, and we both fully understood that people who work with this genus of flies often have little in common. The genus is huge and includes several thousand species, assigned to about a half dozen or more subgenera by taxonomists who study such things. These numbers and assignments change as they learn more, and I have made no effort to keep up with that field. There are people who are passionately interested in the phylogenetic relationships among the species in the genus, but even they tend to focus their attention within smaller, more manageable groupings of closely related species. Then there are the people who work with *Drosophila melanogaster*. Because so much laboratory research has been done using this single species, especially by geneticists, it continues to get the lion's share of attention. When people want fresh material brought in from the wild, they run flytraps that attract lots of *Drosophila* species. The only thing they need to do is separate the *melanogaster* from the other flies in the catch, without bothering to identify all of those other species. With a little practice, this is easy to do. The only problem is that *melanogaster*'s sibling species, *simulans*, also might be in the catch. These two are called "sibling species" because they look alike. The only way to distinguish them morphologically is by a subtle difference in the shape of male claspers (part of the reproductive structures), and it takes a microscope to see whether the claspers are convex or concave. That bit is tricky. What people tend to do is place individual female flies that fit the *melanogaster/simulans* description into separate vials of food and allow them to oviposit. Nearly all wild-caught females will have mated. One then need only wait until their progeny appear to examine the males' clasper shapes to learn whether that wild-caught female was *melanogaster* or *simulans*. There are other examples like this, but that should suffice to explain why one needn't know how to identify all of the fly species caught in a flytrap to get a sample of *Drosophila melanogaster*.

Fortunately, identifying *Biston betularia* by sight is dead easy, compared with sorting out flies. By that point in my career, I was good at it. I had become an expert. But I was not at all expert at identifying all of the other moth species that came to my trap. On that score, I was an amateur. Kitagawa knew this, and he knew that any "log of catches" he asked me to turn in to the forestry service would be essentially worthless. It was, in my view, a case of "form over substance," and I declined to do it. I saw it as fraud. He saw it as satisfying the bureaucracy. Did they need to justify my trapping moths at their facility with a pretense of success, to keep up the appearance of useful work performed?

I might have felt more comfortable with my recalcitrance about keeping a species log had I actually been catching some peppered moths at Takao. But no! When people would ask me, "How many did you catch last night?" and I answered, "None," they were flabbergasted. "None? How could that be?" Often they'd direct me to moths clamped onto a porch ceiling near a light that had been left on overnight. "Lots of moths everywhere!" Indeed there were, but thus far I had caught no peppered moths. My "none" meant "no peppered moths," but they heard that as "no moths of any kind." They thought I was doing a moth species survey. It was hard for me to explain the distinction in my pidgin *Nihongo*.

Kitagawa, to his unstated displeasure, knew that I was unwilling to keep the species log he'd requested, but I also got the sense that he thought perhaps peppered moths were coming to my trap and I might be missing them. Maybe Japanese peppered moths looked different from those found elsewhere in the world, like various races of people do? Fuyama suggested this to me with a broad, friendly smile. I wasn't offended. Indeed, Japanese peppered moths do not look exactly like British typicals: (1) they are duskier and rather similar in shade and pattern to American *cognataria*, and (2) they are somewhat smaller than the British and American subspecies, thus their subspecies name, *parva*. I assured them, however, that I wouldn't miss *parva* in my traps, saying to Fuyama, with my own broad and friendly smile, that the different races of peppered moths around the world looked more like one another than did the diverse races of people—and wherever I go I always recognize people.

They suggested in the most congenial terms that we should visit a museum, where pinned specimens were available for my careful inspection. Actually, I was delighted to accept their invitation to do this. My hope was that the specimens on display would include useful information about *where* in Japan they were caught, and *when*. Different moth species fly at various times throughout the summer, at distinct locations. Sampling their populations required being in the right place at just the right time. The flying interval at any location would only last for a couple of weeks. I desperately needed that information, and my colleagues, as drosophilists, didn't have it about moths. And if Inoue knew about the timing for this species, he didn't offer it during our brief chat.

To be fair, to have such detail at their disposal, even professional lepidopterists rely either on their personal trapping experience for any given species or on access to the trapping records of others. My prompt trapping successes in England and Virginia were due to Cyril Clarke's and David West's years of experience. I had no comparable mentors in Japan.

Unfortunately, the pinned specimens at the museum we visited did not include capture dates or locations that we could decipher. Worse, Inoue's comment about "no known melanics" in Japan was holding up. The specimens we did examine looked very much like the ones I had described, so my Japanese colleagues were at least satisfied that I was competent to recognize their peppered moths, should they come to my traps. The species just wasn't on the wing in Takao at the moment. Japan is a long, mountainous nation, composed of several islands, where the flying times of species would probably correlate with the country's significant latitudinal and altitudinal gradients. With only one summer there—and my late start, to boot—time was running out. I had spent the entire second half of June in Takao. Would peppered moths start flying there in July, or had they already done so during early June, before I arrived? If so, they wouldn't take to the air again there for another year. I'd need to stay in Takao to find out. Or I had to gamble on another location.

31 | Tajima

The train ride from Takoa back to Setagaya and the university involved using the Keio Line, switching trains first at Kitano for Shinjuku, and then at other stations. As I became an old hand at Tokyo train travel, the independence felt good. Kitagawa had drawn detailed maps that served me well. But my next destination, to Fukushima Prefecture (prior to and well away from the nuclear power plant disaster on the coast), would be a bit farther away, staying at Aizu-Tajima Lodge, a lovely retreat then operated by TMU. To get there would require several hours of travel, changing trains along the way. I couldn't carry all of my moth trapping gear as luggage, so I had it shipped by a delivery service. The truck was to arrive there the day after I would. The manager of Tajima Lodge had agreed to meet me at the train station. All I had to do was arrive on time.

This trip would be solo, without an escort. Dr. Tobari was worried I might get lost, so she coached me on how to ride the rails. To start with, we needed to plan my travel schedule, and for that we consulted a calendar. There was a large one hanging behind the door near my desk. It was a Japanese version of a *Playboy* calendar, with a vivid photograph of a stunningly beautiful nude woman on full display above the mundane information we sought (i.e., dates and days of the week). This was 1988, and although spicy pinups were still found on the back walls of auto repair shops in America, they were no longer acceptable in university settings. But I was in Japan, not at William & Mary. I was caught off guard when Dr. Tobari swung open the door to check the dates on the calendar. I did remember it was there, but I was hoping she wouldn't think it was mine! I did not put it there. It was there when I joined the lab. "Not my calendar," I was about to announce, but my self-consciousness faded as she paid absolutely no attention to the photo and focused directly on the upcoming days of the week.

She went on to explain train schedules, what time to be at the station, how to get there, and so on. She would call the Tajima Lodge manager to confirm when he should expect my train's arrival at his end. She also explained the need for me to change trains along the way. She was totally unfazed by the image of the nude woman on the calendar we were both looking at. I was ill at ease and hoped it didn't show.

Early the next morning I headed for the train station, which was a new one to me. I had plenty of time after my arrival to secure my ticket and relax over coffee in the waiting room. When it came time to board the train, I was shocked to see that so many people had already lined up on the platform, before I even thought to go there. Oh no, there might not be room for me on this train! Seasoned travelers know this and don't dither around having coffee. They get in the queue, pronto. Experience is instructive.

Fortunately, we all were able to board the train, and I even managed to find a window seat. As we went farther from Tokyo, I noticed that the Romanji and Katakana characters disappeared, so one had to recognize the station stops in Hiragana or Kanji. Hiragana are long phonetic pronunciations, but Kanji are Japanese versions of Chinese characters. They are complex and woefully alien to Westerners, but once I established a specific search image in my mind's eye for the place where I wanted to go, the exact Kanji popped out of myriad unrecognizable characters and were as easy to recognize as a familiar face. I was becoming a pro—or so I thought.

We arrived at a major station, where lots of people got off and on. I was thinking this might be the place where I should also get off to change trains to go to Tajima. But I hesitated, not sure I had remembered correctly. I fumbled with my notes in desperation as the doors closed and the train pulled away from the station. Well, I'll check out the next stop to see what it is, I decided. What else could I do, pull the emergency cord?

Until that point in my journey, most of the Kanji signs marking the stations along the route corresponded with my abridged schedule, but there were a few minor exceptions not on the list. Surely that was the explanation for the next unlisted stop? Well, let's proceed to another one or two to see if we're still heading in the right direction. That's what I said to myself that day on the train to Tajima.

Then the stop after that was also not on my list. Uh-oh! This became worrisome as unfamiliar Kanji marking the stations along the route mounted. I wasn't quite sure yet what to do, because I still might be on the right train and needed only to be patient until some recognizable Kanji told me where I was. So I sat and stared out the window, hoping for a recognizable sign.

The next stop seemed to be a major one, because everyone got off the train. Everyone but me. I sat there, wondering what to do, still not sure whether I was on the right train. Ah, at last, people were finally getting on. Then I noticed all of these people were wearing blue uniforms, had vacuum cleaners, and were cleaning the coach—a pretty good clue that this was the end of the line.

I went to the open double doors, looked up the tracks forward of the engine, and saw an unmistakable barricade confirming the terminus of this run.

A train conductor came to the door and asked me what I was up to. Why was I not getting off the train? "End of the line?" I asked in my pidgin *Nihongo*. Yes, this was it, I was assured, end of the line. Time to disembark, *kudasai*. I protested, attempting to explain my need to go back in the direction from which I had come. Surely this train would be making a return run, and I wanted to be aboard. "No, no, *iie*," not permitted. I must first leave this train, be processed, and then I could reboard. First things first. "Get off now, please." The conductor was polite but very insistent. Remaining on the train for its return run was not an option.

I fetched my two duffle bags from the overhead bins and moved to the platform. There was only one way to go. Out. But I didn't want to go *out*. I wanted to go *back*. How do I explain that? And to whom?

As I walked to the end of the platform, accompanied by the conductor who'd kicked me off the train, we reached an office of some sort, where other officials spent time. The conductor escorted me in and told those inside that I was confused and needed help. He was right about that. This was the first time my limited grasp of *Nihongo* proved useful. No one there spoke English. Yet they seemed to understand my predicament, probably more from my arm waving than from my spoken words. It was pretty easy for them to comprehend that I'd missed my stop to change trains, and that I could take a train back there to get on the correct train to my destination.

After all, I had a ticket they could easily understand. The hard part was to explain that there was someone waiting for me at the station in Tajima, and I'd be several hours late. I had the phone number for the Tajima Lodge and asked them to call there. They were eager to help me and placed the call, but after a bit of conversation I was informed that the Tajima Lodge manager had already left for the train station and couldn't be reached. It would be another decade before mobile phones would become commonplace, even in Japan.

Finally, at the train station serving Tajima, I was met by the lodge manager himself, not an assigned driver. He greeted me cheerfully and dismissed my late arrival as nothing at all. He made me feel completely at ease as he explained that my not arriving on the previous train prompted him to call his office. By that time, they had received the phone call from the train station personnel at the end of the line, who had explained my missed connection and all the rest. No worries, or so he insisted. His English was better than my *Nihongo*, but mostly we just said how happy we were to see each other.

Upon our arrival at Tajima Lodge, I said "Wow," because there were birch trees planted at the entrance. As it turned out, they were the only birch trees in that vicinity, but their presence was encouraging. We were just in time for supper. I was ushered to a comfortable room, with my own personal hot tub. Sheer elegance.

I could have lived happily forever at Tajima Lodge. It was a beautiful place, and the people who ran it while I was there made me feel like I was the emperor of Japan. Well, maybe not that, but they sure did their best to make me feel welcome. No complaints, except that it has since closed for business.

When my trapping gear arrived by a delivery service the next day, I was hopeful about the prospects of catching peppered moths.

My plan was to spend about a week at Tajima. If peppered moths were there, I'd remain longer. If not, I'd move on. Time was precious, and I didn't know at which locale to spend it. I'd effectively wasted my first two weeks in Tokyo, until I finally began trapping at Takao in the middle of June. Two weeks there brought in literally thousands of moths, but no *Biston*. It was now early July, and I was only just starting on a second location.

I did not ship a gasoline-powered generator with my equipment, for practical and legal reasons, so I relied on several very long extension cords that reached from the lodge to the surrounding woods. I had no trouble in catching numerous moths, with all of the major families represented. Again, there were literally thousands over the course of a few days, many by now familiar me, but still no *Biston*. One afternoon, the lodge manager asked me if I'd like to visit an area where there were a lot of birch trees. He'd noticed my excitement when seeing those few planted at the lodge entrance. He also realized I wasn't happy about the way my trapping was going. I later learned that Fuyama had suggested that my hosts "look at his face" to see if I was happy or sad. This way they could gauge my success, because they wouldn't learn much from my attempts to speak *Nihongo*.

I was so excited to accept an invitation to go for a drive into the countryside, I put on my shoes even before leaving my room, walking all over the tatami. Oops!

So off we went on a pleasant but long drive into the mountains. It was foggy when we arrived at our destination, but it was gorgeous—and covered in birch trees. Surely this was the place I longed to be when running moth traps. "*Denki*?" I asked. "No *denki*," was his answer. The Tajima Lodge manager went on to explain that we were well away from any source for electrical power, so without a portable generator, there was no *denki*.

I could almost kick myself for not having shipped a generator from Tokyo, but we could certainly remedy that by getting another one—albeit expensive—from somewhere. A much bigger problem would be reaching the trapping site from my lodging each day. It was a long way by car, so I'd really need to be living there, among the birch trees, to run my traps. That wasn't possible with my present arrangements. I decided it was time to move on from Tajima Lodge and try another location. A place like the birch woods in the mountains would be nice, if it had both accommodations and *denki*.

32 | Fisheries Lab

Denki among birch woods in the mountains would have been ideal, but that would come later. A little too late for the 1988 season, as it turned out, because first I had to honor the obligations my hosts had taken pains to arrange on my behalf. I was winging it, and so were they. Still, the last thing we wanted to do was offend people who had graciously allowed me to run moth traps at their facilities—and provided accommodations, to boot. After a brief return to Tokyo from Tajima, I was on a train heading for the Tokyo University Fisheries Experiment Station, in a mountainous region of Yamanashi Prefecture.

Fuyama accompanied me on this trip, for some reason. As it was well away from Tokyo and involved switching trains several times, I suspected he was worried I might again become lost, as had happened to me on the much simpler journey to Tajima. "No, no," he insisted, he wasn't babysitting me. He was confident I could find my way. He simply wanted to go along to visit a colleague at the fisheries lab, so this would be a pleasure trip for him. "Otherwise," he chuckled, "you may end up in Tibet."

As we got farther from Tokyo, the style of the trains changed, getting more old fashioned and slower. They also had wide-opened windows, with people hanging out of them. Rather charming, actually. But do trains really go to Tibet from Japan?

The last stop for us still meant a long hike to the fisheries lab. As planned, no one was asked to meet us at the train station, and we ate box lunches after stopping along the path. Fuyama had purchased these earlier in the day. We talked about all sorts of things, including recent books we had read. Fuyama pointed out that many of the those written in English become much longer when translated into Japanese. He cited several then recent examples (e.g., E. O. Wilson's *Sociobiology*, which had expanded from one to four volumes). It takes the Japanese much longer to say the same thing. I didn't argue with this.

Upon arriving at the Fisheries Experiment Station, I expressed some disappointment to Fuyama, since the surrounding vegetation looked even less promising than Tajima. He urged me to give it a few days, to satisfy protocol. Kitagawa had pulled strings to get me there, so please do not embarrass him. Fuyama did not say this so bluntly, but the message was clear: at least put on a show by making a good-faith effort. Sure, why not? I would run moths traps here and hope for the best. There were no options.

Fuyama returned to Tokyo that very night, after getting me settled in. The colleague he went to visit was given the assignment to look after me. I didn't know this at the time. Subtle.

By this point I was an old hand in Japan, accustomed to taking a bath, or a good soak, in a hot tub before or after dinner. I was shown where the baths were, and the sign-up sheet was explained to me. This was a university experiment station, so there were many students there, as well as faculty and independent researchers, of which I was just one. The idea was that people signed up for times to bathe, so it wouldn't become too crowded. I didn't quite catch all of the nuances about how this worked on the first explanation. So, before launching into setting up my moth traps that first night, I signed up for a bath and dove in. It was lovely. I had the place all to myself. It was such a big tub, I could actually swim in it. In fact, I did, thinking to myself, during a lazy backstroke, "Man, this is livin'."

After my luxurious soak and swim, I dried, dressed, and exited the bath area, only to discover a bevy of young women outside the door, all wearing bathrobes and clearly waiting for me to go away. As I came to learn, I was taking my bath during the time allotted to women. The sign-up sheet had a section for men and for women, and I failed to recognize the difference in their various names. It would be like my not distinguishing between a list with entries like Doris, Mary, Nancy, and Susan, versus another one with John, James, Mike, and Bob. I had scheduled Bruce to bathe with Nancy. Nancy, apparently, declined to join me.

After that, they developed a system where they would put a bench in front of the entrance door when it was the women's assigned time in the bath. Or was it put there for the men's time? I wasn't sure about that, even though I asked. Instead, I decided to take my baths early in the morning. This worked out well, as I got up at the crack of dawn to empty my moth traps before the birds got into them. The students seemed to sleep till noon.

We were also on different schedules regarding meals. There was a huge kitchen and dining room at the station, but there were no staff members who prepared meals or cleaned up. The students made their own meals but were pretty lax about cleaning up. Whenever I went into that room, I was a little appalled by the piles of dirty dishes, cooking utensils, and equipment. To use anything meant I had to clean it first, removing the previous user's mess, and then wash it again when I was done. Well, to hell with that. I decided to eat in my room. We weren't on the same schedules in any case, so why should I eat with the students? Besides, by this point during my stay in Japan, I was starting to feel like an *X* in a world of *O*s. I just wanted some private moments.

Not good enough. Fuyama's colleague came to visit me one evening around suppertime. He was clearly worried that I was not enjoying myself. Well, yes, I caught hundreds of moths every night—it's a good place for that—but so far, no peppered moths. Maybe not such a good place for that species, or perhaps the wrong time during the summer? That's what was on my mind. He didn't seem to be bothered about that. He was more concerned that I seemed to be deliberately avoiding interacting with the students. "Why don't you eat with them?" he asked.

After dutifully putting in my week at the Fisheries Experiment Station, it was time to head elsewhere. It was also time to square up my accounts. As the station was not directly affiliated with TMU, I had to pay for what I used, including baths. Fortunately, I was assigned an interpreter to help me communicate with the staff. He was a graduate student from China whose English was better than my Japanese. I knew that right away when we were introduced the first night, as he greeted me by saying, "Good evening, Professor Grant." I asked him "*Denki, wa doko deska*?" to which he responded, "Good evening, Professor Grant."

At checkout time, the manager of the fisheries station asked me several questions, through my collegial Chinese interpreter, about this and that expense, including, "Did you take a bath while you were here?"

"Yes, of course I took a bath while I was here!"

"What *time* did you take this bath?"

Uh-oh! Now I understand his concern—about bathing during the women's time that first night, then late the second night, then early in the mornings after that, when the bath wasn't even officially open for use. I attempted to explain through the interpreter: "I am so sorry. The first night I didn't understand and took my bath at 7. Then, the next night at 10, but after that I . . . "

I scarcely got into my apology before the manager became incredulous: "What, you took *seven* baths on the first night, and *ten* baths on the second night?" Apparently he concluded that I'd taken seventeen baths in two days.

From our conversation I learned that his "*What* time" meant "*How many* times." Good thing I had an interpreter. No wonder the United Nations needs them. Otherwise there might be confusion, possibly leading to war.

"*Ikimashou*," called out the manager. He eagerly drove me to the bus station by the highway, where he dropped me off. He was happy to be rid of me. I don't think he disliked me. He just didn't know what to make of a *gaijin* who ate alone in his room and took so many baths. Getting on that bus back to Tokyo was like going home.

33 | Hokkaido

Back at TMU, sitting at my desk and checking a small pile of accumulated mail, I laughed audibly at a newspaper clipping sent by a William & Mary colleague. It was a cartoon by that creative genius, Gary Larson, showing a woman in a flower-patterned dress that exactly matched her wallpaper, carpet, ceiling, and furniture. Her husband, wearing a striped shirt and checkered trousers, was answering the door. The caption read, "When the monster came, Lola, like the peppered moth and the arctic hare, remained motionless and undetected. Harold, of course, was immediately devoured." I taped Larson's masterpiece to the wall above my desk.

Over the next couple of days, as I prepared for my next trip, students in the lab stopped by to examine the cartoon. They were curious about it, but mostly they just smiled and shook their heads, baffled and not sure what to make of it. Fuyama explained this practice to them, loosely translated as: "When I was a student at UCSD (the University of California, San Diego), I noticed an American custom to put cartoons on the wall. Everybody does it. Their walls are filled with cartoons. I don't get it."

Where to next? That was the question. Kitagawa, again on my behalf, was in touch with people in Nagano Prefecture. I, however, was eager to get north of the main island of Honshu and go to the northernmost island belonging to Japan, Hokkaido. It was my project, and I was technically the boss. But since Kitagawa was making all of the arrangements, in that sense he was. Neither of us wanted to displease the other.

In this instance I won the battle. In retrospect, I would've won the war had I taken Kitagawa's advice and gone to Nagano *then*. That would come later, a little too late, but I was determined to go to Hokkaido, pronto!

I do not regret going there, not for one minute, even though the sample size was $N = 1$ for peppered moths. It was a great experience and ranks among my favorite places visited in Japan. Oddly, when I told friends in Tokyo I was going to Hokkaido, they acted like some Americans do when you tell them you are going to Alaska. Well, I've been to Alaska three times, and Hokkaido only once, and I can report from personal experience that Hokkaido is much more like Honshu Island than Alaska is for any locale in the lower 48. Still, there is a remoteness about Hokkaido that I found especially appealing.

I arrived in the city of Sapporo from Haneda Airport on July 20 and was met by my host, Dr. Masahito Kimura, who jokingly made clear to me that he was not to be confused with "Mr. Neutral Theory," Motoo Kimura. I knew that, of course. Motoo Kimura was a global force in population genetics at that time, while my new colleague, Kimura-san, was a young associate professor of biology at Hokkaido University in Sapporo. He was eager to assist me in any way he could to ensure that my moth-hunting expedition in Hokkaido was successful. He was thoroughly helpful in every conceivable way, arranging for my lodgings and collection trips the whole time I was there. He took a personal interest in the project and worked as hard at it as I did—maybe harder.

Kimura understood my need to be away from the city lights of Sapporo. The morning after my arrival, he took me to an *onsen ryokan*, an unimposing, traditional hot-springs inn near Mikasa in central Hokkaido, where I planned to set up my moth traps. I'd arranged for them to arrive there before I left Tokyo. The manager of the *ryokan* was somewhat concerned about what to feed me and posed the question to Kitagawa by phone. The answer: "Make no special arrangements. Professor Grant will eat anything."

The *ryokan* was situated along a two-lane highway through rural countryside. It was not fancy. Truck drivers stayed in the back row of quarters at the inn. The front section was more upscale, with private rooms and baths. My first night was with the truckers, as my reserved space wasn't yet available. I was in a room shared by several men, who all seemed to smoke in bed. These were cigarettes, not dope. But, smoking in bed? Yikes! They all seemed to look at me rather nonchalantly, but no doubt understood I was not one of them. Their shared toilet was a long pooper—by that I mean, an "outhouse" that was actually inside the building but did not have a flush toilet. Instead, there was a deep pit, below slots in the floor. No doubt the price for staying in this section of the *ryokan* was more economical than a room in the front building, with its own flush toilet. Well, for the first night, I didn't complain, but I did consider taking up smoking.

I was reassigned, with copious apologies, to a private room in the front part of the *ryokan* for the next few weeks. It was luxurious by the previous day's standards. I had round-the-clock hot green tea in a pump thermos, and they provided breakfast and supper in the dining room. *Oishii-deshita*—well, except for the time they served fermented squid. I might have been the only guest in the place to eat that. Even *Nihonjin* turn their noses up at fermented squid. Can't get enough rice to gag it down! But what did I know? So I ate it, much to their amusement. As a reward, the cook promised me potatoes as a special treat for the next night. Lunch was on our own. Over the course of my visit, I got to know the manager/owner/cook. I think I was the first *gaijin* ever to stay there, and perhaps the first he'd ever met. He was curious about me, as I was about him. I'd like to think we became friends. We talked as much as our mutual misunderstandings of each other's language would allow.

The first night at the *ryokan*, I set up my moth traps just above the rear parking lot (behind the truckers' quarters) at the edge of the woods. I avoided going into the deep woods, because I wanted the lights to shine out for at least some distance to catch the attention of passing moths. More practically, that was as far as my extension cords would reach from the source of *denki*. I was alarmed the next morning to see that the lights were out. OUT. What?! Why were the lights out? After barely scrambling into my clothes, I raced to the traps to see what was the matter.

Each of the traps had several dozen moths inside, but none were brimming with specimens, as one might expect on a warm summer night. The mercury vapor bulbs were dead out, and clearly had been that way for most of the night. I soon discovered the cause. The power cord had been chewed through. Clean through! What could have, would have, done that? And, why wasn't whatever did it electrocuted? *Kitsune*. KITSUNE!

Kitsune (pronounced kit-soo-ney) is the Japanese word for fox. These little devils were everywhere in Hokkaido. Unlike American foxes, which hide from humans out of fear for their very lives, *kitsune* enjoy a certain Shinto privilege—they are special in some spiritual sense beyond my reckoning. If a fox wanted to chew through my electrical cord, well, I'd have to learn to work around that.

Surely I wouldn't be allowed to kill all of the foxes around the *ryokan*. Nor did I want to. I actually like foxes. I didn't want to harm a single one. I'd much rather abandon my project than do that. But now what?

I decided to move my trapping site well away from the *ryokan*. I was thinking the fox population might be especially high around the inn, because of human-associated attractions: garbage, mice, etc. So let's increase the distance from all this by going higher up the mountain. *Takai yama*.

There was a primitive road wending its way up the mountain behind the *ryokan*, so I decided to explore that region and try running traps there. It was a great spot for catching huge numbers of moths, well away from other artificial lighting. But by now it was midsummer, and peppered moths were not on the wing. (This is a crucial point, although I was not fully aware of it at this early stage in my career.) Yet, HOLY COW, I'd arrive in the morning to find the lights out and the electrical wires again chewed clean through. *Kitsune*! What's with these little devils?

One night, as I went up to start the generator and provide electrical current for the traps, a fox walked right past me, as if I were a totem pole! That critter didn't regard me as a threat at all. I was flabbergasted by this. So I yelled and hollered and swung my arms up and down to drive it away. The fox left, but not in a big hurry. "Who's that?" it might have thought.

To solve this problem, I hung the extension cords from tree branches, well out of reach of foxes, and I urinated around the generator and the moth traps in places where the wires came back down. I got this idea of marking my territory from Farley Mowat's technique with wolves. It worked for him, but would it work for me? Every night I did this. And—while this is nothing more than anecdote, not science—the foxes no longer chewed through my wires! As a result, I caught a lot of moths, but not a single peppered moth on *takai yama*.

On my way up the mountain each morning, I passed an old woman waiting at a bus stop. She wasn't an odd sight for me, but I was to her. She was in her own country, but I was clearly *gaijin*. She was too polite to say anything to me other than "*Ohayo gozaimasu*," and I returned this standard morning greeting. One day she ventured to ask me what I was doing, but I didn't understand then what she was asking me. After I said my usual "Good morning," I went on my merry way, wondering what I had missed. But her words played on my rudimentary comprehension of Japanese. Ah, so that's what she was asking. Hmm, let's see . . .

As I trudged along my trek up the mountain to unpack the moth traps, I realized the old woman at the bus stop was curious about who I was and what I was doing. Fair enough. Her question deserved the best answer I could provide. I worked on my answer, which I delivered the next morning when I saw the woman at the bus stop. "*Ohayo gozaimasu*," I said, and then launched into the speech I'd carefully prepared the night before to answer her question from the previous morning. It took me a long time to develop my answer, and I'd hoped it'd stick, at least for a few seconds, but nope. As soon as I'd have my say, she'd launch into another series of questions, for which I not only had no answers but didn't even understand what she'd asked. So up the mountain I'd go again, looking forward to what moths I might find in the traps, while wondering what in hell the woman at the bus stop was asking me this time. There was a 24-hour delay between her questions and my answers. Maybe all conversations should be like this?

From time to time Kimura would visit to see how I was getting on. Once he brought me a live peppered moth he had caught on his office window. Hmm? Maybe I should have stayed in Sapporo. Another time he came along with a graduate student and camping gear, so we could try our luck at another location—Daisetsuzan, Japan's largest national park, famous among hikers for its rugged, mountainous terrain surrounding Mount Tomuraushi, Hokkaido's highest peak. This certainly seemed like a good idea, and I very much appreciated the trouble Kimura was willing to go through to help me get a good sample of peppered moths. So off we went, higher into the mountains, on our moth safari. All three of us, including the graduate student who served as our driver, were excited about our upcoming adventure.

After traveling for a short while, we decided to stop at a 7-Eleven convenience store to stock up on snacks for the road. Sometimes people say "7-Eleven" to mean any sort of convenience store that is similar to the familiar 7-Elevens. But this one was, in fact, a 7-Eleven store, by that name. It looked very much like any other 7-Eleven one might encounter in the States, but it was tailored to Japanese tastes. A number of the food items are not likely to be found in American 7-Elevens, especially the snacks. In particular, I had become fond of *kaki-no-tane*, a mixture of crescent-shaped, soy-flavored rice crackers and peanuts. I tried to avoid mixtures additionally seasoned with bits of dried fish skin and seaweed. There were all sorts of things in that 7-Eleven I'd never seen in its American counterpart.

Back in the car, munching away on our snacks, Kimura asked me, "Do you have 7-Eleven in America?"

I was amused by his question. "Yes, we do. In fact, 7-Eleven is an American company, or at least it started as one. It was owned by the Southland Corporation but was recently bought by a Japanese firm." I never paid much attention to the world of big business, so my understanding of these deals has always been vague, but I pressed on. "7-Eleven convenience stores are all over the United States. They used that name to indicate that they had extended business hours, being open from 7 in the morning to 11 at night. These were much longer operating hours than typical grocery stores had at the time. That's why they were called "convenience stores," because they were available at the convenience of the customer. Nowadays most supermarkets have extended hours, and 7-Eleven stores are actually open round the clock, 24 hours a day, 7 days a week. They really are convenient!" Kimura took all this in—or seemed to—by nodding in agreement. Our conversation then took a somber turn when I added: "The problem with staying open all night is that armed robberies happen far too often at convenience stores. For night-shift workers, this is a very dangerous job, especially for women. Now most 7-Eleven stores must hire two people to work the late shifts, even when there are actually too few customers to make this profitable. Regrettably, it's a serious problem."

Kimura weighed all this and then responded, "In all Hokkaido, only one woman can work until midnight."

I repeated what he'd said in my mind, trying to understand what this meant. "In all Hokkaido, only one woman can work until midnight." Who was this woman? I wondered. What did she do? Was she a cop? Was she a hooker? What did she do after midnight? Go home? I repeated Kimura's words in my mind again: "In all Hokkaido, only one woman can work until midnight."

I posed some of these questions to Kimura, until his meaning became clear to me. He was saying that Hokkaido, unlike America, is so safe that a woman—any woman—all by herself, could work all night long without worrying about possible harm. Armed robberies at 7-Eleven stores were unheard of. "In all Hokkaido, only one woman can work until midnight."

Hai, wakarimasu. I do not present this dialogue about 7-Elevens as something unusual. It was typical of nearly all the conversations I had while in Japan, especially when I attempted to speak Japanese. This one, as were most of them, was in English. Every word was clearly pronounced, but the meaning was not at all clear. It took work to figure it out. Once again I marveled at the United Nation's interpreters and was reminded to thank them for keeping peace in the world via their linguistic skills. People who are truly fluent in more than one language amaze me, and I envy them.

I also envied Kimura and his command (albeit less than perfect) of English. That we could communicate at all was entirely to his credit. Left to my abilities with *Nihongo*, we would have been restricted to talking about the weather, how good the food is, or where to find a bus or train station. To have a sophisticated conversation about any subject required the person with whom I was speaking to know English. Otherwise, nothing. So yes, we had misunderstandings, and it took work on both our parts to communicate effectively. A sense of humor helped. Indeed, we actually enjoyed talking to each other. I'm sorry to report, though, that the graduate student grew weary of hearing so much English. He wanted to escape, but he couldn't. We traveled in the same car for hours on end, and he was the driver, while Kimura and I chattered on. We'd break for meals—together—while Kimura and I kept talking. We'd sleep in the same small tent at night, and Kimura and I conversed before drifting off. The graduate student was going nuts. He wanted silence! At the very least, he was sick of hearing English.

Hiro Asami told me the same thing happened to him during his first summer at Mountain Lake. He'd even hear English in his sleep—his dreams were haunted by it. Make it stop! How could he escape this? There was no escape! Welcome to America. I experienced the same feelings at the fisheries lab. I didn't necessarily want to hear English. I just wanted to *not* hear Japanese spoken for at least a good part of my day. Make it stop! So I ate in my room. Alone. And I cherished the solitude. When Kimura explained his graduate student's discomfort to me, I understood it. I might even have apologized to him in Japanese, had I known how.

The graduate student seemed to recover well enough when we pulled into a roadside rest stop. A young woman riding a powerful motorcycle parked right beside us. She took off her helmet and shook her long black hair free, to announce her pleasure at taking a break from the road. Kimura explained to me that the graduate student found her especially attractive. Well, duh! I didn't have to speak Japanese to understand that. In any case, our moods lightened.

That night we settled into a deeply wooded campground. Beautiful mountains surrounded us. It seemed like an ideal place to run moth traps. We probably needed a special permit to trap moths inside the national park, but Kimura seemed unconcerned. Our first night's harvest was huge. But still no peppered moths. Kimura suggested that we should expand our survey by taking advantage of streetlights in the area: along highways, in villages, and across the top of a nearby hydroelectric dam. There were even streetlights in the campground. He had brought along an insect net attached to a long-handled, telescoping extension pole that could reach the streetlight bulbs.

First we practiced in the campground. We'd swing the net through the cloud of moths that were swirling around a lamp high above us. Within seconds the net would be packed with hundreds of moths, sometimes even more. Then we'd bring it down and go through it, hoping to identify moths in the process. We wanted to be able to examine the catch carefully, so capturing too many at a time wasn't efficient. A search image for the correct size and shape made this process easier than it sounds. The three of us worked diligently to examine as many moths as we could, given their level of excitability—and ours. Unfortunately, we had no luck in the campground, despite huge numbers of moths examined by this method.

Next we tried a local village that had a bright mercury vapor street lamp. (Sodium lamps, which produce yellow light—not attractive to insects—were not yet common along roadways.) We went through our routine of sweeping with the net and examining the catch, but apparently our verbal outbursts attracted attention. An elderly gentleman came out into the street, demanding to know what we were doing. Kimura engaged him in a conversation that seemed more like a shouting match. I didn't understand what the man was saying, but Kimura kept barking back, ever more loudly, "Eh?" Over and over again, Kimura said "Eh?" Ultimately we retreated. Once back in the car, I asked Kimura what the problem was. He said he didn't know, as the man's Hokkaido dialect was so thick he couldn't understand a word he was saying. Now, this was news to me. I hadn't realized until that moment that there were different regional accents in Japan. No wonder I had so much trouble understanding *Nihongo*.

Our next sampling attempt was not along an actual street. It was around the lamps that topped a hydroelectric dam. This caper made me seriously nervous. There were regularly spaced mercury vapor lamps on high poles across the top of the dam's walkway, making the whole place look like a formidable prison or a military installation. Kimura brushed aside the idea that we should seek permission, as he knew we'd have reached retirement age before permission was either granted or denied, and he was sure it would be denied for security reasons. Let's just do it and, if caught, apologize. Easy for him to say! But what word should I use to say "I'm sorry"? *Sumimasen*? *Gomenasai*? I wouldn't want to appear feminine.

What worried me was the uniformed guard inside the guardhouse. Was he armed? How would I apologize if he shot me dead? Just how much *did* industrial melanism really mean to me? As we crept closer to the dam, we could see that the lamp posts spanning the dam were too high for us to reach the swirling moths with our net, even with the handle extended to its maximum length. Bummer! "Let's go," I said. Kimura, much more resourceful and daring than I, replied that he thought we could reach the moths at the first lamp, which was right next to the guardhouse, by climbing up on the roof of the building. That'd give us the extra elevation we'd need to catch lots of moths. "Good idea," agreed the graduate student. Good idea?! Did he really think so, or was this social facilitation? As we went up a ladder, we could see through the window that the guard was watching television inside. I still couldn't tell if he was armed or not. I was shaking too much to focus on him, other than to notice that he had his feet up on a table as he sat back, engrossed in a game show of some sort. Kimura told me not to worry about him, but just try not to make any noise.

Silently we moved across the roof, like cats—well, two cats and a nervous dog. Occasionally we'd pause to listen for the sound of the TV, to learn if the guard was still focused on his show. I suppose we hoped he'd turn the sound off if he became suspicious that moth collectors were creeping across the roof. So far, so good. Soon the sound of the water spilling through the dam dominated the air. We even stopped whispering as we swept moths down for examination. Alas, no peppered moths. After a brief spell, which seemed like an eternity to me, we silently crept back across the roof and made our way to the car. Kimura and the graduate student saw this as all in a day's work. My heart was racing and all I could hear was my pulse. Fieldwork!

After Kimura and the graduate student dropped me off at the *ryokan* on their return trip home, I had a few days before I, too, would head back to Sapporo. I spent the time in my usual unproductive activities, but I was impressed by the acts of kindness directed toward me by the staff at the inn. The night before I left, the manager/cook/owner gave me a large tankard of beer at a farewell dinner. He also removed a fanny pack from his waist. Although ubiquitous now, then mainly tourists wore this small pouch on a belt, as a way to hold their valuables. He, for some reason, always wore one and noticed I did not. He insisted that I needed one, especially when travelling in big cities, as I was again about to do. He emptied the pouch of its contents and handed the fanny pack to me, without ceremony. "Here, take it," he might have said. I don't know what his actual words were, but I could not refuse the gesture. As I accepted the gift, I recalled the chilling remark made by my colleague at William & Mary: "Remember, they despise us." Surely he was wrong about that—certainly not all of "them" despised all of "us." I accepted the gift as an act of kindness. What else could it have been? There was no ulterior motive. I was not a diplomat to be catered to. Indeed, it has taken me 30 years to write about this event. I still have that fanny pack, and I wear it whenever traveling abroad. (I use it as a decoy, while hiding my valuables elsewhere on my person.)

I settled into a window seat near the back of an intercity bus heading for Sapporo. It would be a pleasant journey through the countryside, but after about a half hour the bus stopped. I assumed this was a scheduled stop to pick up passengers, but the person who got on raced down the aisle, waving excitedly at me. It took me a moment to recognize her. She was the manager's wife at the *ryokan*—the woman who insisted on doing my laundry, more frequently than I thought necessary. Why was she here? What did she want to tell me? Was there some emergency? Had I forgotten something? Yes, I had. Her outstretched hand held a small Tupperware-like container with a live peppered moth inside. It was the very moth that Kimura brought to me from Sapporo on one of his visits. I had put it in the refrigerator at the inn to keep it alive for a while—not knowing what else to do with it. In any case, I had forgotten about it. It ultimately would have died in the fridge after a few days, or a week or two at most. But now I had it, alive, in the box, on the bus to Sapporo.

At first I thought I should tell this excited woman that she need not have gone through so much trouble, driving halfway to Sapporo and chasing the bus, to deliver this moth to me; that I didn't need this moth for my work; that she could have let it go; that I was so sorry she wasted her time. But before I allowed honesty to ruin her enthusiasm for advancing science, I thanked her profusely for her valiant effort to keep me on track. She then got off the bus and drove back toward Mikasa. Mission accomplished! I and the peppered moth proceeded on to Sapporo, where I released the moth, back at his hometown. Then I met Kimura for supper.

Kimura pulled out all the stops for my sendoff from Hokkaido. He arranged for us to experience a formal tea ceremony. He, of course, had participated in them, but I had not. Why, I wondered, would anyone make a ceremony out of preparing tea? The Japanese do, for reasons beyond my comprehension. I was swept back in time by the ceremony. I won't try to describe it, other than saying that Kimura and I sat on floor cushions as a woman in traditional dress prepared the tea for us to drink. What impressed me far more than the tea ceremony itself was that Kimura arranged to have me experience it. He went through a lot of trouble and expense to do so. I am sure that ceremony, arranged just for me, cost him a bundle of yen, and he was then only at the beginning of his career.

As we said our goodbyes, I told Kimura that my wife was coming to Japan to spend some time with me and would enjoy seeing part of the country. Besides Tokyo, we hoped to visit Kyoto. Kimura had a number of recommendations about what to see and where to stay there. He advised us to also schedule some time in Nara, where he again provided addresses of accommodations. Clearly he knew his country well, and he wanted us to experience it as fully as time would allow.

34 | Tourists

My first day back from Hokkaido was the day of Cathy's arrival in Japan: Saturday, August 13, 1988. I started the day by going to the lab. I had hours to spare, or so I thought, before heading to the airport to meet her. Fuyama was curious about why I hadn't already left for Narita, suggesting that I might already be making a late start. He was right. The trip out took longer than I had expected, and I was getting nervous that Cathy would clear customs and wonder just where I might be. Again, this was before people carried cell phones, and she'd have no way to contact me. I grew impatient with the delays imposed at transfer points between trains and buses, and the long security queues, as I repeatedly checked my watch, scolding myself for not leaving directly for the airport from home.

"Home," whenever I was back in Tokyo, had become the Hotel New Hideoka in Sangengaia. By this time I was a frequent guest there. They knew me. I told the desk staff my *nyohboh* would be joining me for a few days, and they insisted that we move into the bridal suite at no extra charge. Wow, I thought, that was mighty generous of them to give us a larger and fancier room for the price of a single. (Or perhaps I'd been paying too much all along?)

Our timing couldn't have been better, as I arrived at the airport's reception area just as Cathy emerged from customs. From there we worked our way back to Tokyo from Narita, and then on to the hotel to catch up on sleep. Kitagawa and his wife invited us to a welcome dinner that evening at an upscale sushi bar in the heart of Tokyo.

While I had thoroughly enjoyed a steady diet of sushi over the previous two and a half months, I had never seen it prepared like this before. Kitagawa, his wife, Cathy, and I were seated at a long bar, behind which stood the sushi chefs, waiting for our orders. Kitagawa made all of them, one at a time, so we might sample various delights. Behind the chefs there was a long aquarium, in which a variety of fish swam back and forth. As

Kitagawa made his requests, a chef would reach into the water and scoop out a fish, slap in onto a cutting board, slice it up—still very much alive and wriggling—and then serve the perfectly cut up morsels to us, sometimes on a dollop of rice and sometimes not, sashimi style. Each presentation was made with a flourish. Talk about fresh! And, of course, there was *biru* aplenty. *Gochisosama deshita.*

Mrs. Kitagawa spoke very little English, so she remained mostly quiet and smiled, while the rest of us chatted amiably. Kitagawa was charming, as always, and asked Cathy about herself, her family, her job, her impression of Japan, and so on. He asked me why I hadn't caught any moths in Hokkaido. Well, we did catch at least 50,000 moths, I assured him, but only one of them was a *Biston betularia parva*, the one at Kimura's window in Sapporo. "Now are you ready to go to Nagano?" Kitagawa asked.

Agreed! I would go to Nagano after Cathy returned to Virginia. But until then we would become tourists.

I did want to have another brief go at Takao, to see if perhaps *Biston* might be flying. Because it was only a short train ride there from Tokyo, we could tour shrines, palaces, and gardens by day and still have an opportunity to run moth traps at night in the national park. Cathy enjoyed (or so she said) assisting my moth-trapping efforts in Takao, including refueling the generator in the middle of the night. She was a bit leery of our climbing over the perimeter wall after dark, although she was not nearly as worried about being caught by the police as encountering venomous snakes.

Two such species are fairly common in that park. One is a pit viper called *mamushi*, which looks somewhat like our familiar copperheads but perhaps is a bit duller in color. I hadn't seen any on my rounds. The other, called *yamakagashi*, is conspicuously marked with contrasting blocks of yellow and red on dark olive or black scales. I saw these several times and even pointed one out to Cathy on her first visit during daylight hours. At night, we deliberately stamped our feet as we tramped along the trails through the woods, to warn nearby snakes of our approach, in the hope that they'd retreat. Being bitten is far more likely if you accidentally step on one.

During our daytime visits, I also pointed out Japanese giant hornets, the largest, most venomous, aggressive hornets in the entire world! Early on, I noticed the people who worked in that national forest all wore white hardhats, not so much as protection from falling tree limbs, as to hide their black hair. Whenever a giant hornet flew into view, the foresters would dive to the ground, fearful of being stung. Fuyama explained that the hornets had an especially powerful toxin to drive bears (*kuma*) away from their nests. Bears love to eat hornet larvae, and hornets attack bears on sight. As most Japanese have black hair, it seems the hornets mistake *Nihonjin* for *kuma*. Ergo, the white hats, which announced to the hornets, "I am not a bear!" Neither Cathy nor I have black hair, but as a precaution we, too, wore white hats in the woods. Still, we were occasionally buzzed by hornets that seemed to be checking us out. Inspection flights?

Within a few days it was clear that *Biston* was not on the wing then at Takao, so I packed up my traps, generator, and gas can and hauled everything back to the lab at TMU, to bundle it all up for the next destination. Cathy and I carried the lot by commuter train. I can't picture being permitted to do that today. Imagine a gasoline-powered generator and its fuel container as hand luggage on a train. Yet we did it without anyone paying attention to us. Once back at the lab, however, I was confronted by a graduate student, who had been assigned by Kitagawa to retrieve us, along with our gear, by car. Somehow I hadn't gotten word about this arrangement, and the graduate student was very worried that Kitagawa would blame him for letting us down. I attempted to reassure the student that it was perfectly all right, but he was not so sure how Kitagawa would regard his failure to carry out orders.

We spent the next several days and nights sampling the sights, sounds, and foods of Tokyo—including touring the Imperial Palace, gardens, shrines, and the bustling shopping districts of Ginza, Shinjuku, and Shibuya. On the way back to Sangengaia on one of those evenings, we spotted a McDonald's restaurant. Today, of course, they are ubiquitous, but that was our first such sighting in Tokyo, and we wondered if it was like the McDonald's we might find back home. After a steady diet of Japanese cuisine, we decided to give it a try, more out of curiosity than from a Big Mac attack. Sure enough, we couldn't tell the difference in the fare, except that it was more difficult to order food in Japan. Whatever I asked for in any restaurant always prompted offers of numerous options. Here, they were the sort of familiar questions you'd get at any American McDonald's, such as "Do you want fries with that?" or "Do you want cheese on that?" or "Would you like a large?" Except these were in Japanese and were spoken so rapidly that I had no idea what they were asking. My strategy was to simply repeat my order until they finally gave up offering options and served me what I asked for. I would ultimately win these contests of politeness, but it was a struggle. At this McDonald's, however, they were determined to please me, no matter how hard I tried to get only what I ordered. After the first server failed to have me accept any of the options he suggested, he responded to my insistent cries of "*Wakarimasen*" by calling in reinforcements. From the back, out came my server's supervisor (or so I suspected, from his demeanor). He figuratively clapped his hands together, as if to ask, "What seems to be the problem?" Whatever else he did say was in Japanese. Round two. I ordered the same items, and the supervisor offered me the same options—perhaps even more options, but still in Japanese. It was a standoff, but we were hungry by this time and I accepted the lot: cheese, fries, large drinks, the works! It was clear why that guy was the supervisor.

After our late supper, followed by a train ride back to our hotel, Cathy discovered she no longer had her purse with her. "What?! Are you sure?!" She was sure. What she wasn't certain of was the last place where she remembered having it with her. We had had a busy day and visited lots of places. The last one was McDonald's. Did she leave it there, we wondered? Well, it was now quite late at night, and the fast-food restaurant was hours away. Surely it would be closed by the time we could return there to look for her purse. We decided to go back first thing in the morning, to be waiting when they opened. In the meantime, we'd try to sleep, worrying about the whereabouts of the purse and its contents: money, traveler's checks, airline tickets, credit cards, car keys, and Cathy's passport and driver's license.

I could tell from her tossing and turning during the night that Cathy was having a hard time getting to sleep, but when I woke up bright and early the next morning, she was finally sawing wood. I was used to getting up at the crack of dawn to check moth traps, so I went back to McDonald's alone. They weren't open yet when I arrived, but I could see employees inside getting ready for business. I tapped on the counter window by the street. That got the attention of the daytime manager. He slid open the window and looked at me curiously. I greeted him with "*Ohayo gozaimasu*" and a selection of apology words (*sumimasen* and *gomenasai*, without regard to gender), as if I'd stepped on his foot. Then I attempted to explain to him, in my rehearsed pidgin *Nihongo*, that my wife (*watakushi no kanai*) may have left her *saifu* ("purse," or I may have said "suitcase") inside the restaurant last night. Without batting an eye, the manager asked me in perfect English, "What is wife's name?"

As soon as I told him, he wrote it on a piece of paper, closed the window, and disappeared from sight. I had paced on the sidewalk for a few minutes when suddenly the window slid open again. The manager's head popped out and he asked, "Is this wife's purse?" Sure enough, it was. I thanked him in every language I knew as I took possession of it. He nodded, pulled his head inside, and slid the window closed. They still weren't ready to open for business and he, no doubt, had work to do.

By the time I returned to the hotel, Cathy was awake and dressed. She exclaimed, "You got it!" when I handed her the purse. Well, yes, but I didn't yet know if everything was still in it. She searched through it very carefully. It was clear that her wallet had been opened, probably when the manager at McDonald's checked for the owner's name, but it was also clear that nothing was missing. Absolutely everything was there, much to our great relief, if not surprise. *Arigato gozaimashita.*

Put riding the bullet train on your bucket list. The *shinkansen* was unlike any train I had ever traveled on previously in Japan—or anywhere else, then or since. It doesn't go everywhere in Japan, as it requires special tracks. But where it does go, it gets there *fast.* Fortunately for us, it went from Tokyo to Kyoto, a distance of 476 km, in under two and a half hours. Swoosh! Speed records have been broken since we were there, and they continue to be updated, but even in 1988 the experience was extraordinary!

Inside the plush coaches there are large speedometers on the bulkheads, so passengers can marvel at how fast they are going, since the countryside, viewed from the windows, is a blur. Well, only if you stare straight out. The scenery is still enjoyable if you look forward as it rushes toward you. The sensation is peculiar, because the ride is so smooth it's hard to believe the velocity at which you're traveling. Attendants serve food and beverages from carts pushed along the aisles, as they do on airplanes, yet—no matter what the speed—the liquid in the drinks doesn't even shimmer. As Fuyama had advised, there was no room for me to bring along my moth traps, so we headed to Kyoto without field equipment. We were unencumbered tourists.

Kyoto, although a large city of about 1.5 million, is very much smaller than the sprawling 13.5 million in metropolitan Tokyo. Before Tokyo became the capital city of Japan, Kyoto had that honor for a millennium. Tokyo, once called Edo, had its name changed when it was promoted. "To-kyo" is a rearrangement of the syllables of "Kyo-to." For those interested in early Japanese architecture and gardens, Kyoto is not to be missed. Better still, if that's possible, is Nara, an even more ancient capital of Japan, earlier than Kyoto.

Cathy and I enjoyed our visits to Kyoto and Nara immensely, with special thanks to Dr. Kimura for his tips on what to see and where to stay. As the end of August approached, it became time for Cathy's airplane to head back to America, and my train to Nagano.

35 | Nagano

Ten years before the Winter Olympics were held in Nagano, I went there to trap peppered moths. Although I had no clue then that Nagano would become a worldwide household name within a decade, I did know it was a favored spot for Japanese skiers and for those who loved a good soak in the natural hot springs. It was, and remains, just about the most popular tourist destination in Japan. For good reason.

I was met at the train station by Dr. Katsura Beppu, who (like his colleague, Dr. Masahito Kimura, in Hokkaido) was another drosophilist. In fact, just about all of my hosts throughout Japan were drosophilists—some geneticists, some ecologists, some taxonomists—who all shared a common interest in this incredible dipteran genus with our distinguished colleague, Professor Osamu Kitagawa. Dr. Beppu was on the faculty at Shinshu University in Nagano. It has several campuses, and he was the director of the Institute of Nature Education in Shiga Heights, where I was to stay and run moths traps. When I arrived, students were still in residence at the field station, so I spent my first night at a very posh resort hotel, with local hot springs that fed the baths. It was pure luxury. Too bad Cathy missed this place! Who but kings get to live like this? Fieldwork can be so demanding!

The next day I had the Education Center all to myself. The students had returned home, and only Beppu popped in and out, as his schedule would permit. During the day I spent long hours at a desk, correcting manuscripts in an attempt to make them understandable. This, in fact, was a difficult job, and it took a lot of time and serious concentration. To my Japanese colleagues who would ask, "Please check the English," it may have seemed like a trivial thing for me to do. The problem wasn't finding English words, it was first discovering what it was the authors were trying to say. I struggled hard with this, as so many of their sentences made no sense to me. So many came across like the puzzling utterance, "In all Hokkaido only one woman can work until midnight." I figured that one out because Kimura, who had said it, was there with me in person, and that allowed me to question him until his intended meaning became clear. While reading manuscripts, the authors weren't present, so I had to make guesses. That proved to be difficult, and sometimes I guessed incorrectly. I still recall trying to figure out what one author meant when he described "brandishing a glass slide." Yikes!

It's easy to make jokes about such odd choices of words, but imagine yourself (assuming you are not Japanese) trying to write a paper in Japanese. I know I could not do it. I wouldn't even attempt it. Although I am officially certified (from my PhD degree program) as having a reading knowledge of two foreign languages—a requirement back in the 1960s—I wouldn't begin to write formal papers in either of them. Yet some scholars routinely publish in several tongues. Theodosius Dobzhansky did. So did Ernst Mayr. I tip my hat to them. Most Americans of my generation don't—or can't.

So why do my Japanese colleagues insist on writing papers in English? Easy answer. Today the international language in science (as well as many other fields) is English. If authors want to reach a global audience, they must write up their work in English before it can be published in the major journals, especially *Nature* and *Science*. Scientists around the world, whatever their native language, must publish in English or perish in obscurity. Americans are spared this struggle. How lucky for us! A letter from Kitagawa accompanying one of his manuscripts invited me to "Please change my Japanese English into 'elegant' one." His request was one of many I had during my stay in Japan. In Nagano, Beppu added another. So that's how I spent my daytime hours—rewriting papers into "elegant" English. At night I brandished moth traps.

I knew from the minute I looked around at the landscape that Nagano was *Biston betularia* country! Maybe it was the birch trees that suggested that to me. I could've kicked myself for not going there straight away. So many places to try, so little time. Yet, while it looked like ideal peppered moth habitat, it was still well away from Japanese industry. First things first. Let's see if I can at least catch peppered moths in Nagano.

You can run moths traps until hell freezes over, and if you don't catch a particular species, you cannot say with certainty that it isn't there. You can only note that you didn't catch any. But, by golly, if you do catch one at a certain location, you can say—with absolute certainty—that it is there. Absence of evidence versus evidence of absence, blah, blah, blah! Which is better? Evidence of presence trumps all! It is flat out qualitative. Nothing is so definitive. Do peppered moths live in Nagano? Yes! How do I know? I caught one at the Shindai Field Station on September 7, 1988. I might attempt to describe here how I felt at that moment, but I defer to a much better writer. E. O. Wilson—in his charming memoir, *Naturalist*, published in 1994—described how he felt about finding an ant species he was hunting for on one of his expeditions. I know that feeling. I think all field biologists do.

Beppu was happy for me, and he kept the specimen alive in his fridge, just in case we might catch a female to breed it to. We were pretty sure the season was all but finished, as I also found a mature *Biston betularia* caterpillar in the vicinity. It was in the final instar stage and had clearly been feeding for a few weeks before I found it. It should pupate within days, but by that time I'd be on my way back to Virginia. I wasn't about to smuggle it home, so I suggested to Beppu that he feed it fresh leaves for a few days until it pupated. Then, in the spring, it'd eclose and he could confirm whether it was a *Biston* and score its adult phenotype. That was the plan as I packed up my personal effects to leave Shiga Kogen, Nagano's ski resort area, for my journey home.

I left my last two moth traps with Beppu, so he could continue trapping until the season was over, although we were about at that point already. He could also resume trapping in the spring, as he promised to do. More practically, he could use them to introduce students to the joys of moth trapping. *Biston* or not, there were lots of moths in the area, which his environmental students could learn to trap and identify. I had also left one of my three *gomibaku* traps with Kimura in Hokkaido, for the same reason. I would have no further use for them, and I had a good supply of traps back home. Getting students interested in moth surveys appealed to all of us.

Oddly enough, I was satisfied about the brief time I spent in Nagano. A beautiful part of the world, by any standard. In addition, I found what I was looking for there: peppered moths. Well, hardly a sample size to boast about, yet a qualitative indicator that this was a place in which to spend some time. My colleagues would do exactly that. It was a locale well away from industry, so, based on the traditional account of industrial melanism, Nagano would hardly be the place to expect to find *melanic* variants of peppered moths. But at this stage of my incipient Japan investigation, I was just happy to find the species I came to study. We made plans.

Plans, by definition, are for the future. At that point, it was time to get me back to Tokyo for my return to the United States. But first we would take in a little sightseeing in the region. The Jigokudani Snow Monkey Park was a blast. Beppu took me there, just for fun. This is a place every nature lover should visit. What are commonly called "snow monkeys" are Japanese macaques (*Macaca fuscata*). They live in large troops, ruled by a dominant male, whose status is advertised by an erect tail. He prances around, daring any subordinate to defy him. I know if I were a snow monkey, I'd want to kick his butt. Then again, if I did, I'd prance around until someone kicked mine. Oy vey! How human they are!

What really amused me was how the monkeys all lollygagged in the thermal pools. They just sat around in water up to their eyeballs, as if they were in a sauna at a posh hotel. "Hey! We're relaxing here!" In winter they depend on that heat. They soak themselves in the hot baths, even while snow is thickly piled on top of their heads! Now, in early September, were they just relieving stress or something? I couldn't resist asking Beppu, "Did the monkeys learn to do this by observing Japanese people soaking themselves in hot tubs, or are the Japanese imitating their monkeys?" I'd have thought that Beppu had heard this query many times before, but he had no answer. "Good question," he said.

Beppu took me to a nearby village to visit his parents. We were to have lunch before he put me on the train for the journey back to Tokyo. I marveled at the immaculate beauty of his parents' home and gardens. I had seen magazine photos of swept gravel gardens, producing geometric patterns of serene beauty, but seeing the real thing at a private home was new to me. At first I reacted politely, as anyone should have, with "Yes, isn't that nice." But they left me out there for a while as Beppu's mother prepared lunch. For a brief time I was alone in the gravel garden—although it was a significant period by Japan's crowded standards, with people everywhere. Rake tracks in the gravel went here and there, up and down, but as my solitude sank in, so, too, did the patterns of the tracks. I actually can't explain this, nor shall I try. I was just somehow caught up in it. I was moved in some strange way, one I do not understand.

After a lunch beautifully prepared and served by his mother, Beppu brought me two large balls of yarn or thread. One was the size of a softball, and the other was a bit bigger. Yarn? Thread? I'm not exactly sure, even now. The smaller of the two was woven or sewn into multiple geometric patterns that produced kaleidoscopic likenesses of chrysanthemums. The larger one was more impressionistic, with the faces of cranes. Wow! was all I could say when I saw these. I was totally blown away by the sheer beauty and creativity that went into this art form. Beppu then said, "My mother wants to give you one of her creations. Please choose the one you would prefer."

"Uh-oh," I said to myself. What I expressed to Beppu seemed to me, at the time, to be diplomatic: "To choose one is to reject the other. Both are beautiful. I would prefer it if your mother decides."

I now have BOTH. Each sits on a pedestal, on a mantelpiece above a fireplace in my house: a place of honor where our treasures are displayed. I certainly was not trying to coax, move, or persuade Beppu's mother to give me *two* of her works of art. Even having one was more of an honor than I deserved or imagined. I was just trying to be nice—and politely tactful. I failed miserably, and I put Beppu's mother in a position where she thought her only way forward was to give me both of her *temari*, instead of just one. I have since learned that these incredibly beautiful and intricate balls of thread are presented as gifts to distinguished visitors. I was regarded as such by Beppu's mother. She offered me one of her exquisite *temari*. I took two.

36 | East Meets West

Seppuku, or *harakiri*, fortunately is not a mandatory penalty for failing to sample peppered moth populations in *Nihon*. Yet I was deflated by having put so many *Nihonjin* to so much trouble to advance my project. I tried very hard to succeed, but didn't. How was I to describe my *haji* in the seminar Kitagawa invited me to deliver to his department before I left Tokyo?

I spent days preparing my talk. I organized and reorganized my slides. These were two-by-two inch, mounted 35 mm transparencies, in the standard carousel format in use everywhere before PowerPoint was invented. I had brought some of those slides with me, and others I had taken while in Japan. I had a straightforward report on the research Rory Howlett and I had published in a British journal earlier that same year. The article was about morph-specific behavior (or lack thereof) in background selection by peppered moths, and the exciting (to evolutionists), ongoing, solidly documented declines in melanism at that time in Britain, which I was personally privileged to observe and record over the previous several years as Sir Cyril Clarke's protégé in Liverpool. The message I had hoped to bring out in my seminar was that we were now actively searching for parallel evolution in other industrialized nations. Ergo, the very reason for my coming to Japan. I was hoping to enlist Japanese colleagues in this quest. To be charming and diplomatic—and maybe to show off—I intended to present this seminar entirely in Japanese. I spent a lot of time writing it out, word by word, in *Nihongo*. After a while, I realized I was more concerned with how it sounded than with what it said. Defeated, once again, I decided to give my talk in English, with Japanese thrown in here and there, in the hope of keeping everyone on board. I even invited people to ask me—in Japanese—to repeat whatever they might have missed because of my poor command of their language. As it turned out, I depended on them. My language skills did not rise above gibberish. Fortunately, my seminar audience was tolerant. They smiled patiently and applauded at the end of my talk, much relieved that it was over. Then we went out for beer.

The epilogue to the Japan saga is that Hiro Asami saved my butt.

After he finished his PhD at the University of Virginia, Hiro went back to Japan and picked up the peppered moth population survey in his spare time. He, of course, had other fish to fry as a new assistant professor, under the gun to produce publishable research in his own program, which involved directional coiling in snail shells and its potential role in speciation. Hiro came out of graduate school with both feet on the ground and the energy to move them productively.

Ultimately we published the results of our joint study in the *Journal of the Lepidopterists' Society* in 1995, with the title "Melanism has not evolved in Japanese *Biston betularia* (Geometridae)." We based that assertion on a grand total of 307 peppered moths, trapped at nine locations over three years. Seven of the 16 sites that were sampled yielded none of this species, and one location in Nagano, sampled in 1992 and 1993, provided 68% of the peppered moths caught. Although I wrote the paper (since it was in English), I insisted that Hiro's name go first, simply because the bulk of the data were his. Without his follow-up work, there would have been no publication at all. Plus, Hiro recruited a number of his colleagues to supply collection data from other places in Japan. All of the moths were the typical phenotype—that is, of the 307 from these nine locations, there was not a single melanic in the lot. It appeared at the time that Hiroshi Inoue was right—there were no melanic peppered moths in Japan. The weakness in our study was that our collections of this species were only successful well away from Japan's industrial centers. We really wouldn't expect to find melanics—certainly not at appreciable frequencies—in unpolluted habitats. We hadn't worked the country as thoroughly as would be required to draw firm conclusions.

Nevertheless, Hiro and I felt an obligation to report what we did find from the surveys we had made. Basically, we corroborated—with real data—Inoue's statement that melanic peppered moths were unknown in Japan. Years later, however, that assertion was contradicted.

In 2000, I attended the European Congress of Lepidopterology, held in Białowieża, Poland. Although I am not a lepidopterist, by giving a talk at such a gathering, I was hoping I might recruit expert moth collectors who'd be willing to keep an eye out for melanic peppered moths in Eastern Europe. The meeting was very well attended, with participants from all over Europe and places beyond.

When it comes to speaking foreign languages, Europeans put Americans to shame. While talks were permitted and delivered in other languages, by far most of them were in English, despite the fact that few attendees were from English-speaking countries. Lucky for me! I would not have understood a presentation delivered in any other language. There were Germans there who could speak French, Danes who could speak Italian, and so on. Yet no matter where they were from, or what other languages besides their own they might also know, every last one of those attending this conference could speak English. Again, lucky for me!

What they couldn't do, however, was speak Polish. Western Europeans admirably master the languages of other *Western* European countries, but Slavic languages are given short shrift. This conference was in Poland, so the host lepidopterists were Polish. As educated professionals, they, too, were fluent in English, so they communicated with their guests from other countries very nicely in that language. The receptionists at the hotel desk also were fluent in English. Things went smoothly. But some members of the local staff did not speak English particularly well, or perhaps at all. This was rarely a problem, but I took the opportunity to show off when one of my colleagues seemed to be having trouble getting the attention of a waitress. He wanted more beer, and he was barking at passing waitresses in English. So I chimed in: "*Przepraszam, Pani! Dwa piwo prosze*." She soon returned with the two beers I'd ordered, and I thanked her, "*Dziekuje*," to which she responded "*Prosze bardzo*."

Somewhat astonished, my colleague said: "You speak Polish?"

Rather than tell him my Polish grandmother babysat me during World War II, I said: "Of course I speak Polish. Doesn't everyone?"

I barely got that out of my mouth when a Japanese couple attending the conference approached our table. What incredible timing! This was just too good to resist! Without batting an eye, I greeted them in Japanese and invited them to join us. Our exchange in Japanese lasted only a few seconds, but it was enough for my European colleagues to exclaim, incredulously, "You also speak Japanese!"

By this time they were flat-out flabbergasted that I, an American, could speak Polish and Japanese! Little did they know that they had heard just about my entire vocabulary in both languages, but I couldn't resist asking mockingly: "Don't you Europeans teach foreign languages in your schools?"

I opened my talk with a review of the classic tale of industrial melanism, no doubt familiar to all in the lecture hall. I then focused on the recent declines underway in England, especially well documented by Cyril Clarke, and the work we had been doing in North America in the twelve years since I had been in Japan. Could changes like those we had seen in England, America, and parts of Western Europe also be occurring in Central and Eastern Europe, and across China? That was the point of the talk. I was eagerly trying to recruit colleagues to find out. As these meetings were being held from May 29 to June 2, it seemed an ideal time to get an early sample of peppered moths. The night before my talk, I ran an improvised moth trap on the balcony of my hotel room and caught about a half dozen of them with barely any effort. They were all typicals, and gorgeous. I put them in a hanging net (cloth screen) cage and brought them to the lecture hall to display to the audience during my talk.

I began my presentation by confessing to a lecture hall full of professional lepidopterists that I was not one of them. My interest in peppered moths in particular was the role that species played as the classic example of natural selection that could be observed in real time, and the well-documented history of genetic changes in its natural populations. I was attracted to study peppered moths as an evolutionary geneticist, just as I had once worked with *Drosophila* fly species and a couple of parasitoid wasp species. If you want to watch evolution as it happens, pay close attention to insects! I felt it necessary to reassure this learned audience that I was not pretending to be a lepidopterist by presuming to give a talk at their conference. I came to them because I needed their help.

Somewhere along the line, I held up the hanging cage and said, "So, if you are interested in keeping an eye out for peppered moths, I brought some along so you might see what they look like." Stony silence met this unintended insult, so I quickly added, "Well, in truth, I brought these specimens along so you'd see that I know what they look like." That got a great response—my biggest laugh of the lecture.

Very near the end of my talk, after spending most of it on the data from other countries, I mentioned, almost as an aside, that we failed to find melanic peppered moths in Japan. During the brief post-talk discussion period, Dr. Soichiro Kinoshita, the gentleman I had welcomed to our beer table the previous day, raised his hand. When I called on him, he rose to his feet and announced: "Melanic peppered moths do occur in Japan. I caught one. I'll send you my data."

Instead of data, Dr. Kinoshita sent me photographs. One photo showed the typical (pale form, or wild type—*Oo-shimofuri-eda-shaku*) Japanese peppered moth, *Biston betularia parva*. The other was a fully melanic variant of that same species, easily as black as a British *carbonaria* or American *swettaria*. It was the real deal. With Dr. Kinoshita's kind permission, I published his photos in a 2004 *Journal of Heredity* article on the genetics of melanism, focusing primarily on American and British peppered moths. In that paper I stated, "With the exception of a single melanic specimen captured by Dr. Soichiro Kinoshita, . . . Japanese peppered moth populations are monomorphic for the typical phenotype" (p. 98). As I type this now, 30 years have passed since Asami and I asked Japanese moth trappers to keep an eye out for melanic peppered moths. So far, none—beyond Kinoshita's sole specimen—have come to our attention. No doubt others probably exist. But without data to the contrary, there is no reason to suspect that melanics have become common anywhere in Japan.

PART VI

37 | Serendipity

A silver lining emerging from the absence of melanic polymorphism in Japanese peppered moths was that it provided the naïve genetic background we needed to test the hypothesis that the genetic dominance in phenotypic expression of the *carbonaria* allele evolved over time.

Theories about the evolution of dominance date back to the founding of population genetics (in the late 1920s) and vary, depending on their authors. This subject has been reviewed by Denis Bourguet and won't be repeated here. The particular theory supported by Kettlewell's hybridization experiments (published in 1965) argued that when conditions lead to the selective sweep of a new mutation, modifiers that promote its expression (dominance) over the allele it is replacing are also favored by selection. Thus a new mutation, not initially dominant, becomes so as a result of the accumulation of background modifiers.

Kettlewell was attracted to this theory, because he noticed that the earliest museum specimens of melanic peppered moths were not as dark as those caught in more-recent years. This was consistent with the theory that dominance evolved. It was also in accord with the argument that these early specimens might have carried different alleles (e.g., those for f. *insularia*) or simply had faded from loss of scales over the century since their capture—a familiar problem with preserved material. Kettlewell needed to do more. And he did. A lot more!

Kettlewell reasoned that if a modern, fully dominant allele for melanism were placed on a suitable genetic background where modifiers had not accumulated, dominance would break down. If not, then this dominance was a physiological property of the allele—an argument proposed by Sewall Wright in 1934. To answer this question, Kettlewell hybridized British melanics and North American *Biston betularia cognataria* caught in an area of Canada where melanic polymorphism was unknown. His assumption was that because melanism had been historically absent in that part of Canada, natural selection would not have accumulated modifiers promoting the dominance of absent alleles—that is, the Canadian stock he chose would provide the requisite naïve genetic background to test the evolution of dominance theory.

Kettlewell backcrossed melanic progeny from the intercontinental F_1 hybrids to the Canadian stock and continued backcrossing for several more generations, thus increasing the proportion of Canadian genes to British genes by 50% with each generation. Kettlewell reported a dramatic breakdown in the melanic phenotype within three backcross generations, and he published extensive photos of the progeny each step of the way. This was convincing evidence that the dominance of the melanic allele in British peppered moths evolved by the accumulation of background modifiers and was not a direct physiological property of the allele.

As a matter of routine, science assesses the correctness of work by testing its repeatability. Sooner or later, all claims that have any substance will be reexamined. The breakdown in dominance from Kettlewell's hybridization experiments stood unchallenged for 11 years—until 1977, when David West reported no such breakdown in his hybridizations between British and North American moths from southern Virginia. But West's caveat was that melanic polymorphism did exist in Virginia populations, albeit at a low frequency, and he further demonstrated, through controlled crosses, that the American melanic allele was also dominant to its common wild allele. Thus West's failure to corroborate Kettlewell's support for the evolution of dominance was easily attributed to not placing melanic genes onto suitably naïve genetic backgrounds.

Seven years after, in 1984, Kauri Mikkola also reported no breakdown in dominance using *Biston betularius* (= *betularia*) typicals from Finland for his hybridization studies with British melanics. Because melanism was unknown in Finland, it potentially provided an appropriate naïve genetic background. Yet melanism, especially in f. *insularia* (whose alleles are dominant to the *typica* allele), is known elsewhere in Europe, so one could argue that sufficient gene flow could supply modifiers recurrently, even if they are not maintained there by selection. Of course, that same argument could be made about Kettlewell's choice of Canadian stock. Melanic polymorphism occurs elsewhere in North America, and Canadian populations are not closed to gene flow.

Kettlewell's claim that the dominance of the *carbonaria* allele evolved as a result of an accumulation of modifiers had been challenged, but those attempts also had deficiencies. What was needed was a source of peppered moths where (1) melanic polymorphism did not occur and (2) was closed to recurrent gene flow from where it did exist. Japanese *Biston betularia parva* jumped out at us from the Far East.

Again, it was Hiro Asami to the rescue! Hiro provided the stocks of *parva* from females he'd caught in Japan. From there, Cyril Clarke and I divided the lab work. He would attempt to repeat Kettlewell's experiments using British melanics (*carbonaria*) hybridized to *parva*, and I, through hybridizations and backcrosses, would place American melanic genes onto *parva*'s naïve genetic background. We were off to the races!

At only one generation per year, races using peppered moths move at a glacial pace, compared with working with fruit flies. For starters, to get a breeding stock of captive moths, one must catch females, and most moths that come to the traps are male. But every so often we get lucky. Taking the eggs a mated female lays, we must feed their developing larvae for most of the summer and into autumn, when the mature caterpillars pupate. These pupae are kept cool (overwinter) until the spring, when they eclose, emerging from the pupal stage into adults. Then they can be mated, and the mated females produce eggs that hatch into larvae, which must be fed, and so on. So, from the time females are first brought in from the wild, two years will have passed before F_1 hybrids are available to be backcrossed to a parental stock. If all goes well, the first backcross generation appears the year after that. To carry on this process of backcrossing until the gene of interest is placed on a genetic background that is primarily (93.75%) derived from only one of the two parental subspecies requires three generations of backcrossing beyond the initial hybridization producing the F_1 generation. To arrive at where Kettlewell got would take five years, assuming no difficulties occur.

Rarely does all go well when working with peppered moths in captivity. Feeding large numbers of voracious larvae is vastly time consuming, keeping pupae healthy during the winter requires paying close attention to how they are stored and to their overall condition, and coordinating the emergences of adults to use in making specific crosses takes careful timing—as well as a lot of luck. Moreover, getting the right combinations of matings to be successful is pretty iffy. Sure, peppered moths mate in captivity, but some don't, and too often these are the very crosses necessary for the study. Even when eggs are produced from the desired crosses, they don't always hatch. Still, we do have some successes, and we rejoice in them.

Then there are times when this work abruptly stops, because of a loss of the necessary breeding stock that would allow it to go forward. That's the big bugaboo. To start all over after a few generations is daunting. Anyone attempting to repeat Kettlewell's work in its entirety had better be determined in thinking it worth all that effort. It is not a trivial undertaking, nor a minor investment of time.

So, yes, we were off to the races!

Kettlewell described the typical phenotype of the American subspecies, *Biston betularia cognataria*, as more closely resembling the intermediate form of British *insularia* than British typicals. Sir Cyril Clarke described Japanese *B. betularia parva* as resembling typical American *cognataria* in their wing pattern, though perhaps a bit paler. Lady Féo Clarke referred to *parva* as a "boring little moth."

Boring or otherwise, American and Japanese peppered moths look more alike than either does to the British subspecies. But such generalizations overlook the fact that there is considerable geographic variation in the appearance of moths classified as typicals *within* the same subspecies. What is f. *typica* in one part of North America doesn't necessarily look like typicals found elsewhere. Unfortunately, no formal study of geographic variation among f. *typica* has been made to date. Even typicals caught at the same locale show some variation. From my own observations as I traveled around various regions collecting peppered moths—explicitly looking for melanics—I couldn't help noticing differences in the paleness of typicals in disparate parts of the country. The typicals I trapped in the Upper Peninsula of Michigan were lighter in shade than those I trapped in the southern Appalachians of Virginia. To be sure, there was plenty of overlap, but some of the specimens in the birch stands of the far north could easily pass for British typicals—at least in paleness, if not pattern. The relevance of this became apparent as we pressed on with intercontinental hybridizations of British *betularia* to *parva* and *cognataria.*

The formal statistical analyses of the data from these hybridization experiments were published in the *Journal of Heredity* in 2004. In short, there was no evidence of a breakdown of dominance of American melanic alleles hybridized with Japanese *parva*, and then backcrossed for one generation beyond the F_1 (i.e., full dominance expressed on a 75% naïve background). Backcrossing beyond that generation stopped after the loss of breeding stock. Cyril Clarke's work halted even sooner, getting only to the F_1. Yet a clue to explaining Kettlewell's interpretation of a breakdown in dominance came from other hybridization experiments with an entirely different objective: a test for allelism between the American and British melanic genes.

38 | Allelic Melanism

The genetic basis for melanic phenotypes in British peppered moths has been as firmly established as any character can be by Mendelian crosses. It was worked out through the analysis of 12,569 progeny, obtained from 83 separate broods reported by numerous independent workers. The heritability of melanism in this species is beyond dispute, except by those who willfully ignore the published evidence.

Furthermore, the genetic basis for melanic phenotypes in the American subspecies of peppered moths, *Biston betularia cognataria*, was demonstrated from Mendelian crosses by David West.

As the full melanics on both sides of the Atlantic (f. *carbonaria* and f. *swettaria*) are inherited as autosomal (any chromosome other than a sex chromosome) dominants, it was assumed that they are alleles at the same gene locus. But are they?

To answer this question using a Mendelian genetics approach requires making crosses between British *carbonaria* and American *swettaria* to perform a test for allelism.

How does this work? Well, the classic test for allelism involving two *recessive* mutations is to cross the two mutant forms and examine their progeny. If the progeny are also mutant in phenotype, then the genes involved are alleles at the same locus. If, however, the progeny are normal (or wild type), the relevant genes are not alleles at the same site. Diagrams may be helpful:

Case 1, the two recessive mutants are alleles at the same locus:

a^1a^1 (mutant) × a^2a^2 (mutant) → a^1a^2 (mutant)

Case 2, the two recessive mutants are at different loci:

aaBB (mutant) × *AAbb* (mutant) → *AaBb* (normal wild-type)

In Case 2, what was wrong with *aa* was corrected by the *A* allele carried by the *bb* parent, and what was wrong with *bb* was corrected by the *B* allele carried by the *aa* parent. In other words, they complemented each other.

By way of analogy, suppose two cars of the same make and model won't start. That's the mutant phenotype: *won't start*. But maybe they won't start for different reasons. One car may have a dead battery, and the second, a bad starter motor. We can recombine parts from these two cars and get a car that *will start* by exchanging batteries. (Exchanging starter motors would also work, but that would be a harder job for the mechanic.) If both cars had the same thing wrong with them—say, dead batteries—exchanging batteries would still result in the same phenotype: won't start. Two cars that won't start for the same reason are analogous to two organisms that are defective because they have mutant alleles at the same locus. Two cars that won't start for different reasons are analogous to two organisms that have the same or similar phenotypes that are caused by mutations at different loci. The bottom line is this: when two mutants are *recessive*, the test for allelism is direct and dead easy.

When the two mutants are *dominant* genes, however, the test for allelism is neither direct nor easy. This is precisely the case for British *carbonaria* and American *swettaria*. The same phenotypic results are predicted among the progeny in the F_1 generation from crosses between *carbonaria* and *swettaria*, whether their genes are alleles or at different loci. A one-locus model and a two-locus model predict the same *phenotypic* ratios in the first generation, but different *genotypic* ratios.

A cross between a heterozygous American melanic and a heterozygous British melanic under the one-locus model can be diagrammed as follows:

$$Cc \times C'c' \rightarrow \tfrac{1}{4}\ CC' : \tfrac{1}{4}\ Cc' : \tfrac{1}{4}\ C'c : \tfrac{1}{4}\ cc' \text{ (3:1 phenotypic ratio)}$$

Under the two-locus model (using different letters), however, the diagram would be:

$$Aabb \times a'a'B'b' \rightarrow \tfrac{1}{4}\ Aa'B'b : \tfrac{1}{4}\ Aa'bb' : \tfrac{1}{4}\ aa'B'b : \tfrac{1}{4}\ aa'bb' \text{ (3:1 phenotypic ratio)}$$

Here we should note that the heterozygosity of the two parents can be confirmed. If either parent was homozygous for the dominant melanic gene, all progeny in the first generation would be melanic—that is, the appearance of typicals among the progeny would confirm the heterozygosity of both parents. It is from these heterozygous parents that we predict a 3:1 phenotypic ratio of melanics to typicals under *both* the one-locus and two-locus models, so from the F_1 generation, no distinction can be made between them. What is clearly different is the genotypic distribution predicted by each model. These distinctions are hidden from direct observation until these progeny can be test crossed to determine their genotypes. That requires another round of crosses and an additional year of work.

In all, I attempted 17 testcrosses of F_1 intercontinental hybrid melanics to their siblings of the recessive (*typica*) phenotype. Of these, 14 produced broods, of which 10 resulted in scorable adult progeny two years after the initial hybridization between the subspecies.

The key outcome expected from the one-locus model is that one-third of the testcross broods should yield melanic progeny only. Three of the 10 successful broods were indeed melanic only, and of these, one had a family of 65 progeny—all black! Under the two-locus model, no melanic-only broods are predicted. From this alone, assuming an independent assortment of genes on nonhomologous chromosomes, the two-locus model is falsified.

The progeny in the remaining broods included both melanics and typicals, as expected in both the one- and two-locus models. For the one-locus model, these are predicted to occur in 1:1 ratios. For the two-locus model, either 1:1 or 3:1 ratios are predicted from segregation at one or both loci, respectively. The qualitative distinction between the one- and two-locus model hypotheses was the observation of all-melanic broods among the testcrosses of the F_1 generation, including one very large brood. So, while a tight linkage remains a lingering possibility, we can reasonably conclude from these Mendelian results that *carbonaria* and *swettaria* are alleles at the same gene locus.

More recently, Ilik Saccheri's group in Liverpool discovered a transposon (a DNA segment that can change its location) inserted into the cortex gene (which influences wing-color-pattern development in other lepidopteran species) of all British f. *carbonaria* they examined. The transposon was absent in typicals and *insularia*. Since very dark *insularia*, barely distinguishable from fully melanic *carbonaria*, do not possess similar insertions in their cortex genes, this suggests that a conspicuously darkened phenotype is not transposon dependent. Curiously, the authors did not discuss this in their 2016 report. Perhaps future work might also include a molecular examination of the American melanic, f. *swettaria*?

Before leaving this subject, a discussion regarding the appearance of intermediates among the testcross progeny is perhaps relevant in reassessing Kettlewell's evidence supporting the evolution of dominance. British typicals and American typicals differ in appearance, not only in their patterning features, but especially in their degrees of darkness, with the American specimens being smokier in color than their British counterparts. This variation showed up in the testcrosses. The ratios of smoky intermediates *combined* with the paler typicals, compared with full melanics, conformed to 1:1 Mendelian predictions from such crosses. An *insularia* allele was clearly not involved in these hybridizations. Otherwise, none of the F_1 progeny would have shown the typical phenotype. Instead, the intermediates observed among the testcross progeny in the test for allelism appeared to be variations of the *typica* phenotype, resulting from the recombination of genes underlying phenotypic differences between the British typical (*betularia*) and the American typical (*cognataria*). Kettlewell described *cognataria* by its phenotypic similarity to middle-rank British *insularia*. Surely intermediates, such as those that conspicuously appeared in my testcrosses for allelism, must have also appeared in Kettlewell's backcrosses of hybrids to Canadian stock. Perhaps it was these intermediates that he interpreted as demonstrating a breakdown in dominance of the *carbonaria* allele? We may never know for sure, but that's my guess for now.

If there is anything to the theory that dominance evolves through the accumulation of background modifiers, support for it does not flow easily from work on melanic peppered moths. Instead, the combined evidence to date suggests that dominance is a property of the allele.

39 | Conspecific Pheromones

In 1991, I hand-delivered a large number of *Biston betularia cognataria* pupae to Sir Cyril Clarke in England. He was curious about whether females from the American subspecies could attract British males to assembling traps by their pheromones. The hope of a *quantitative* comparison between the effectiveness of American and British females as lures was obviated by the paucity of British females Cyril had available that summer. His emerged from a small number of pupae kindly provided by Mr. Tony Liebert, an esteemed amateur lepidopterist. American females, thanks to the hefty stash of pupae I'd smuggled into England, were in abundance, so we managed a convincing *qualitative* test. That is, do American females attract British males, or don't they? The answer was conspicuously unambiguous.

On the very first morning—when we opened the assembling trap and it was teeming with British males, lured there by the pheromones released by American females—we had a clear answer. I said as much to Lady Clarke as we excitedly unpacked the trap. "Yes," she replied, "We have our answer." Cyril added, "We'll write a paper." And we did.

Indeed, that first morning kicked off a tremendously successful season, when a combined total of 933 British males were caught, of which more than half were attracted to the assembling trap containing American females. Compared with the grand total of 154 moths caught in the Clarkes' garden the previous year, 1991 was outstanding for us.

Cyril enjoyed comparing the attraction between American and British peppered moths to the attraction between American servicemen and British women during World War II, but Lady Clarke persuaded him not to put this in our report.

Another paper was foremost on Cyril's mind when I returned to West Kirby that summer. It was about to appear in the prestigious *Proceedings of The*

Royal Society B. It would be about the lost colony of scarlet tiger moths (*Callimorpha* [*Panaxia*] *dominula*). In 1961, Cyril's long-time colleague, Philip Sheppard, established a colony of scarlet tiger moths on a railroad bank outside their normal range by releasing approximately 13,000 tiny caterpillars there, the progeny from controlled crosses that were expected to produce known proportions of three adult forms (phenotypes) called *typical*, *medionigra*, and *bimacula*. (Again, "typical"—as used here—is the name applied to the common form that resembles the original type specimen used by taxonomists to describe a particular species.) After nearly 30 generations in West Kirby, the proportions had not changed, in marked contrast to changes recorded in the parent population from Cothill (near Oxford).

Cyril learned of the experimental colony's survival in 1989, when a stray scarlet tiger moth happened into his mercury vapor lamp trap—odd for a day-flying moth. After some investigation, the colony was discovered to be thriving along Wirral Way, a popular walking trail along a former railway bank not far from the Clarkes' residence at 43 Caldy Road.

Cyril was positively bubbling with excitement about the discovery of the lost colony. "Would you like to see them?" he asked with such eagerness that I could hardly say no. We all squeezed into his shiny black Mini, both he and Lady Clarke packing walking sticks, with Cyril at the wheel, and off we went to arrive at steep outdoor stairs that would take us to the trail leading to the *dominula* colony.

While the details of the scarlet tiger moth story never seized my attention, I was delighted it gave Cyril such pleasure. No kid opening presents at Christmas could smile as broadly as Cyril Clarke with a living scarlet tiger moth in the palm of his hand or perched on his shoulder, like a pirate with a pet parrot. The part of the story that became interesting for me was sparked later that summer, when I received a reprint of the 1991 *Panaxia dominula* paper. Another would follow later, from *Oikos*, published in 1993. Both had Denis Owen as a coauthor.

40 | Howard Hughes Lecture

Why do we need another symphony? There are plenty of them, and good ones, too. Some I haven't even heard yet. Only a philistine would seriously ask such a question. There is always room for more good music, although what qualifies as "good" is in the ear of the listener. Ask mountain climbers why they want to scale a peak, and the customary answer is, "Because it's there."

Yet I have been asked—not infrequently—to explain the value of my research. "What good does it do anybody?" Indeed, the husband of my sister-in-law posed that very question more than once. Each time, I explained to him that my research brings me personal enjoyment, like playing the banjo—or, more grandiloquently, similar to how a composer might feel when writing music, or a rock climber's thrill during an ascent. For me, the primary driving force is curiosity. While the ultimate goal of basic research is to describe the universe, practicing scientists tend to carve out smaller pieces to explore.

"But will your research lead to anything useful?" is among the more insistent queries I get, to which I answer with another question: "Who knows?" I am not being facetious. Though the goals of basic and applied research are different, they are not mutually exclusive. This was the reason I invited Sir Cyril A. Clarke to present the Howard Hughes Lecture to the William & Mary community.

One of the annual events at the university is a contest called the Raft Debate. Students choose the contestants from among the faculty. The idea is that three people are adrift on an imaginary raft, with sufficient provisions for only one. Two of the individuals must be pitched overboard, to allow the other to live. But who should survive? And how is this decided?

Each of the contestants represents one of three broad areas of study:

Area I, the humanities (e.g., art, literature, music, philosophy); Area II, the social sciences (e.g., economics, government, history, sociology); and Area III, the natural sciences (e.g., biology, chemistry, geology, physics). The survivor, in theory, carries into the future the collective knowledge from only one area, while the store of knowledge from the other two areas would be lost at sea. Which area, then, is most important to humankind? Ergo, the debate. Each contestant explains to the audience why his or her area is most important and should be saved. They also take this as an opportunity to dump on the other areas—just for laughs. There is also a devil's advocate, who argues that none of these areas is worthwhile, so all three on the raft should be pushed into the drink, leaving no survivors. At the end of the debate, the audience picks the winner by applause. A moderator, rather than an applause meter, gauges crowd enthusiasm and declares the winner. No booing!

I was honored one year to represent the natural sciences. When later asked how I'd done, I boasted: "I won! I came in second!" (Success for scientists was not coming in *last*. Our fan base was in the lab, rather than in the audience. Right!) My appeal to save Area III was built on the simple fact that scientists are people, too. Scientists, as bona fide human beings, also care about art and music and (some) literature, so allowing the scientist to live was more likely to preserve a broader slice of humanity than is probable by saving someone who gleefully eschews science. So far, so good. I was on a roll in this role. I fleshed out my argument by providing a famous example: "Einstein played the violin."

My opponent from the music department shot back, "If you ever heard Einstein play the violin, you'd know why he became a scientist."

Bazinga!

We were pleased when the applause was hearty for all of us. The music professor and I seemed to be tied, requiring a two-way clap-off to reveal the winner. Unluckily for me, the student orchestra was in the audience, and they knew how to orchestrate a crescendo.

The Raft Debate also revealed the tendency for nonscientists to conflate discovery and exploitation. To persuade the audience that the scientist deserved to be pitched into the drink, my opponents brought up loathsome *applications* of science, starting with the atomic bomb. But what about the good things science has done? The Raft Debate aside, the list is very long.

Most people making such a compendium would probably fill it with medical and technological advances, rather than, say, an understanding of bird songs, or genetic polymorphism. What I thought might be enlightening for the William & Mary community would be to "eavesdrop" on the musings of Philip Sheppard and Cyril Clarke regarding genetic interactions. Their basic research focused on polymorphic color patterns and wing-shape variation in African swallowtail butterflies (*Papilio dardanus*), with the color and shape characters controlled by genes at different loci. Extending their curiosity to different nonallelic interactions in other species led them to a medical breakthrough: a therapy preventing Rh hemolytic disease, a lethal genetic condition in human babies. They did so by unraveling clinical records, which revealed significant interactions of Rh genotypes and A, B, and O blood-type incompatibilities (see Chapter 11). Although Sheppard had died since then, Cyril Clarke was very much alive, and he could still tell the story—straight from the horse's mouth.

I invited him to come to William & Mary to deliver the Howard Hughes Lecture. To sweeten the pot, we arranged for him to receive an honorary degree. He accepted without hesitation, chuckling that he already had accumulated nine honorary degrees but was hoping to get into double figures.

We arranged accommodations for Sir Cyril and Lady Clarke, along with the other recipients of honorary degrees to be awarded on that Charter Day in February 1992, at the posh Williamsburg Inn. They stayed in Williamsburg for about a week, during which time I chauffeured them around to see sites in the historic area. It was a particular part of history in which both of them had considerable interest. Cyril felt the place should still belong to England and didn't hesitate to say so, but always with a disarming chuckle.

On one of our tours around Colonial Williamsburg, we were greeted by an interpreter at the Governor's Palace, who asked our group if there were any visitors from other countries. Cyril raised his hand. "Which country, sir?"

"England," he replied.

"When did you arrive?" she inquired.

"Yesterday," said he.

"Oh my," she went on. "Are you experiencing any jet lag?"

"No. I never get jet lag." Cyril then explained why, raising his wrist to display his watch: "The secret is I never reset my watch when traveling. I keep it on Prime Meridian. That way I needn't bother with local time."

Our drive around the Yorktown battlefield was a more somber excursion, once Cyril and Lady Clarke understood that this was the very location where the British forces under General Cornwallis surrendered to George Washington in 1781. Normally full of chatter at the sites, Cyril merely managed to say "Yes" from time to time as he examined the redoubts.

We hosted a dinner party, so the Clarkes and several of our faculty friends could get acquainted. Cathy did all the hard work by preparing the food, while I served as bartender. We had a grand time. Lady Clarke was full of compliments, ranging from the poise of our daughter Elspeth, who was studying ballet, to the elegant meal that appeared out of nowhere. The latter was in praise of Cathy's efficiency as a gracious hostess. Cyril, as usual, was charming—to perfection, much to the delight of our guests. Entertaining them was easy for us, all the way from drinks to dinner to dessert. It was going so well.

Then Cyril glanced at his wristwatch, at first casually, then shocked, as he boomed out, "Look at the time!"

It was 2 o'clock in the morning in Greenwich, England. Our dinner party thus ended at 9 PM in Williamsburg, Virginia, as hastily as if someone had shouted "Fire!" in a crowded theater.

The next day I took Cyril and Lady Clarke to Millington Auditorium, so he could review the slides he'd prepared for use in his lecture, "Butterflies and Babies." This would give him a feeling for the room and would allow him to have a bit of practice. Very fortunately, Lady Clarke was there to remind him of what some of the slides were meant to display. I grew worried about whether Cyril was up to the job. By now he was at an advanced age and seemed startled when some of the slides came up. He'd impatiently demand,

"What's that?" Lady Clarke would tell him. Uh-oh, I thought. This could be bad, but I kept that concern to myself.

I had invited a number of people from other universities to Cyril's lecture, including David West from Virginia Tech, Jim and Bess Murray from UVA, and Lincoln Brower. He sent regrets, because of a scheduling conflict, but conveyed his warmest wishes to his old friends. David West also asked me to invite Doris Zallen, a science historian. This proved to be most fortunate, as Doris later interviewed Cyril and provided him with some missing information about Philip Sheppard's railway bank colony of scarlet tiger moths. In addition, I advertised the talk widely at William & Mary and sent invitations to nearby universities in the region.

Lady Clarke was pleasantly amused to call Cyril's attention to a huge banner over the auditorium entrance that read, HEAR SIR CYRIL CLARKE. "Hear," she repeated. "Not see, or listen to, but *hear* Cyril Clarke."

At 4 PM I began my introduction of Cyril by asking Lady Clarke to stand and be recognized by the audience, saying, "They really are a team." After the lecture, Doris Zallen thanked me for acknowledging Lady Clarke's role and reinforced that it was both true and appropriate.

Once Cyril began his lecture, all of my concerns about his being up to the job evaporated. He was on! He reveled in it. Millington Auditorium was packed. There were people sitting on the steps along the aisles. They laughed at his every humorous remark, and there were many. He was a consummate performer and didn't miss a beat—or misidentify a slide. I occasionally glanced at Lady Clarke, and her eyes were riveted on her husband. He finished to thunderous applause. (Surely he'd have won the Raft Debate!)

Following his talk, our department chairman, Larry Wiseman, presented Cyril with a William & Mary Biology Department T-shirt that Larry had personally silk screened. Without the slightest hesitation, Cyril removed his suit coat and replaced it with the T-shirt. "Now I am dressed for dinner." The audience roared its approval.

The dinner was for the recipients of honorary degrees on Charter Day, including their invited guests, as well as the obligatory bevy of notables such august occasions demand. It was such a huge celebration it had to be held in the basketball field house. As we entered the arena, a portly man in a tuxedo greeted us with a broad grin and asked David West, conspicuously the tallest person in the crowd, if he played basketball. Dave, smiling equally broadly, had a ready reply: "No. Do you play miniature golf?" The portly man laughed and said, "Touché." We were all in a jovial mood for the soirée. We were also hoping Cyril forgot to wear his wristwatch.

Joining our table was a local newspaper reporter, Frank Shatz. He asked me to arrange an interview for him with Cyril Clarke. Mr. Shatz was a holocaust survivor. Cyril had served as ship's surgeon with the Royal Navy during World War II and was assigned the task of giving Winston Churchill his physical exam. Cyril and Mr. Shatz had war stories to share.

The president of the university capped the evening with an after-dinner speech. Cyril seemed much relieved that the Colonies appeared to be in good hands. He later wrote to the university president to express his approval.

I decided to scrub the Clarkes' planned flight from Newport News to Dulles Airport for their return to England. In as much time as the first leg would take, between the layover and the waiting, I could drive them. It would also save them the needless jostling and confusion of getting between terminals at Dulles. As their flight wasn't until late in the day, I arranged a visit to the Smithsonian Institution to see the Lepidoptera collections. I contacted the director/curator, Dr. John Burns, and he arranged for the Clarkes to meet the museum's staff. They rolled out the red carpet.

Part of the sweet deal was that they would permit me to park in the lot reserved for staff members. Great! I was a frequent visitor to the Smithsonian and a card-carrying member. Nevertheless, normally I had to park elsewhere. But now I was escorting VIPs!

So off we went, early in the morning. Along I-64, heading west from Williamsburg, Cyril commented on how many trees there were. Sure, England has trees, but nothing like what we have in Virginia. Then he saw a highway sign indicating the distances to Richmond and Charlottesville. He read these aloud and wondered about Charlottesville, pronouncing the "Ch" as in Charles, more "char" (as in *char*coal) rather than a softer "shar." He mused constantly about such things. His mind was always at work, and nothing escaped his attention.

Then up I-95 we went, into Arlington on I-395, zipping past the Pentagon, and arriving in Washington, DC, America's bustling seat of government. Lady Clarke was impressed with how flawlessly I drove and how well I knew my way around such a confusing place, especially with so much traffic. She dubbed me a "genius." Ah shucks, I'm not exactly in that class.

Then we arrived at the Smithsonian, and I was on the lookout for the parking lot entrance I was told to take. Somehow I went right past it. I decided I'd just go around the block and try another shot at it. I turned right at the corner. The road I entered went into a tunnel and emerged into heavy traffic a long way from the Smithsonian. I knew something was amiss when I drove past the Pentagon for a second time, all to the chatter of Lady Clarke telling Cyril what a genius I was. Well, maybe they won't recognize the Pentagon, or perhaps they might think we have two?

They clearly enjoyed their visit to the Smithsonian and the fuss the staff made over them. I thank John Burns for making that possible. But all good things must come to an end, so it was back into the car for our trip to Dulles Airport. Now it was rush hour, and my driving genius was put to the test. Every time I go there by car, I marvel at the people who commute on a daily basis. I was so happy I chose to live where I could walk to work.

At Dulles we were supposed to be met by a representative of the airline, who would look after the Clarkes until boarding time. She did not show up. So I undertook her role and ushered the Clarkes into the airline's VIP lounge. It was my first time in such a place. Wow! How pampered some passengers are. Money buys a lot. They had all sorts of food to choose from, and unlimited drinks. They even had single malt whisky. Cyril said, "Have a big drink." I declined, explaining that I still had to drive back to Williamsburg. Even a genius shouldn't drink before driving. But I enjoyed a snack with them, and we talked.

One of our topics was, What comes next? Cyril brought up Japan. He thought I should go there. I reminded him that I'd already done that, and Hiro Asami was still on the job. "Yes," said Cyril.

Then I asked him about Denis Owen. Cyril had invited him to collaborate on the reassessment of the polymorphism in Philip Sheppard's railway bank colony of scarlet tiger moths, as Denis was a recognized expert on scoring the several phenotypes in this species. Their coauthored papers (one already published, with another to follow) offered and drew much criticism about variability in scoring *medionigra*, but I paid scant attention to that particular controversy. What interested me was the identity of the co-author: Denis Owen.

I knew that was the name of the person who, 30 years previously, documented the rise in melanism in American *Biston betularia cognataria*, paralleling the rise in the British subspecies. I wanted to meet this guy! I wanted to ask him questions. But *not* about scarlet tiger moths. Cyril agreed to arrange our introduction.

41 | Mr. Parallel Evolution

Denis Owen had a long, varied, remarkably productive career, beginning well before he entered the zoology program at the University of Oxford under the mentorship of the renowned ornithologist, David Lack. By the time Denis completed his undergraduate studies in 1958, he had already published over 40 research papers. Who else has done that? From Oxford, he went on to the University of Michigan to earn his PhD. It was a fortunate choice, for it was there, on the outskirts of Detroit at the Edwin S. George Reserve, a field research station operated by the university, that Denis trapped American peppered moths in large numbers. He did this in his spare time while working on owls for his PhD thesis.

Denis began trapping American peppered moths on the George Reserve in 1959, coincidentally the very same year Cyril Clarke began trapping British peppered moths near Liverpool. The difference between their projects is that Cyril Clarke continued trapping at the same location every year until his death in 2000, with his devoted assistants squeezing out a few more years of sampling, whereas Denis Owen ran traps in Michigan for only three years, discontinuing the study after he completed his PhD and moved to Africa in 1962, to accept his first academic position at Makerere University in Uganda.

During those first three years, from 1959 to 1961, the percentages of fully melanic peppered moths in populations near Detroit, Michigan, and near Liverpool, England, were nearly identical. Clarke and Owen were unaware of each other when they began their work. Clarke started trapping moths in his backyard garden, at the urging of his friend and colleague, Bernard Kettlewell. Owen was thoroughly knowledgeable about Kettlewell's extensive publications on industrial melanism, but he began trapping moths independently, motivated by his unquenchable curiosity. When 90% of the American peppered moths attracted to his 100-watt mercury vapor lamp were melanic, he immediately recognized the importance of his discovery: *parallel evolution.*

Young Denis Owen didn't just sit on his data. He knew what he saw! It was abundantly clear—from 515 among the 576 *cognataria* he trapped—that the melanic phenotype of American peppered moths overwhelmingly was the common form near Detroit, Michigan. Indeed, Owen included in his survey several other species, in three separate genera, from nearby counties outside the George Reserve. The data clearly demonstrated that melanism in moths was a general phenomenon in southeastern Michigan, not isolated to a sole species at a single location.

Were black moths always common there? Or had their numbers increased to that high frequency over time, as had been documented in the industrial regions of England? Before 1848, melanic peppered moths were unknown in the United Kingdom. All specimens captured by moth collectors were of the typical (pale) phenotype. But within about 50 years (and moth generations) after the first melanic specimen was discovered, melanics had all but replaced the typicals, reaching frequencies of 98% near Manchester. In 1959, what Denis Owen discovered was that the frequency of melanic peppered moths in southeastern Michigan matched their frequency near Britain's conurbations. In the United Kingdom, though, there was a trail—published documentation of the rise in frequency of melanism, recorded by moth collectors. What was the history of these moths in America? Without knowing that, Denis simply had data that showed 90% of the peppered moths in southeastern Michigan were black. Was that parallel evolution, or merely a coincidence?

Denis needed more data from previous years. Yet how do you go back in time? Museums! They are worth their weight in gold in preserving history. Historians have always known that, but what about biologists? To understand what is happening now, one must know what was happening then.

Denis Owen demonstrated the enormous value of insect collections by inspecting preserved moth specimens kept in natural history museums across North America, in order to discover exactly when on the calendar melanic peppered moths were first officially recorded in various regions.

To visit distant museums requires travel, lodging, and time. All of that demands money. Graduate students, then as now, are notoriously skinned. Established scholars spend a lot of time pursuing grants to support their research programs. So much has been written on this topic that I'll pass on adding more, other than to say that Denis Owen exhibited remarkable resourcefulness in securing the funding he needed to get the job done.

And he *did* get the job done. He recorded the dates when the very first melanic specimens were placed in museum collections in various geographic regions of North America. Earlier collections included only the "typical" (i.e., matching the type specimens) representatives of their species. Owen did not argue that melanics were absent in natural populations prior to those dates, but he did assert very effectively that moth collectors—not unlike stamp and coin collectors—enjoy finding the unusual. If anything, Owen contended, it is the rare form that tends to be overrepresented in collections. It is hardly likely that an experienced moth collector would ignore or discard a rare form, but almost certainly would keep it and record it. His point was this: the first recorded specimens among museum collections probably reflect fairly accurately when rare forms started to become common enough to be caught by collectors. Prior to those dates, we might safely assume that the form was rare, or at least not common.

What Owen showed from his painstaking analysis was that melanic peppered moths were first recorded in North America in 1906, near Philadelphia. Melanics then began popping up here and there across the industrial northeast since that time, but they did not nose into the Detroit region until 1929. In general, Owen concluded, the spread of melanism in North America began about 50 years later than in Britain.

42 | Aerogrammes

My most vivid recollection from 1993 was an aerogramme from Cyril Clarke declaring it "imperative" that someone go to Michigan "at once" to assess the current level of melanism in *cognataria*. The melanic form of peppered moths at his location in England was then 25%, down about 40% from when I started collaborating with Cyril only a decade ago—f. *carbonaria* was taking a nosedive! "What is happening in America?" Cyril wanted to know.

He was not directing *me* to find out. That was not his style. He asked if I might know someone who could trap moths for us. I agreed that assessing melanism in Michigan must be done without delay and wrote back that I would go to Michigan the very next summer. While I had contemplated doing that for some time, it took Cyril's gentle nudge to not put this off any longer.

Following that significant correspondence, I contacted the University of Michigan about the possibility of running moth traps on their Edwin S. George Reserve in Pinckney. I was referred to Richard Alexander, then the director of the Museum of Zoology. Alexander was well known for his extensive work in insect behavior and, as it turned out, was an old chum of Denis Owen. He seemed pleased to accept my proposal and put me in touch with the then director of the George Reserve, Ronald Nussbaum, who was also the curator of amphibians and reptiles at the Museum of Zoology. After a brief correspondence, I was instructed to contact the museum's administrative secretary, who would work out the details for my visit.

Not long before I left for the George Reserve, an aerogramme arrived from none other than Denis Owen himself. First contact. Very formal. He said Cyril Clarke had informed him that I was going to revisit his trapping sites on the George Reserve, and he thought it would be "imperative" (that word again) that he and I put the traps in precisely the same locations, so temporal comparisons would not be confounded by trap placement. Good idea! I would be positively delighted if Denis would join me. Plus, he might compare what the place looked like when he was there in 1959 with how it appeared 35 years later.

We soon switched from aerogrammes to faxing to stay in touch. In a fax to Denis sent June 22, 1994, I said, "I'm leaving for Michigan next Tuesday (June 28)." Comparing that date with the combined catch records from the George Reserve during 1959–61, spanning the periods May 14 to August 31 each season, it is patently obvious that we didn't plan well.

When I organized my first trip to the George Reserve, I was unaware of Denis's detailed trapping records, and he had all but forgotten about them. Three decades had passed since Denis donated the specimens he'd caught to the University of Michigan's Museum of Zoology, during which time he had taken a faculty position in Africa, had written several books (including one on tropical butterflies), transferred to a faculty post in Sweden, and ultimately returned to England to join the faculty at Oxford Brookes University, where he directed the projects of numerous graduate students and continued his own research on birds, moths, butterflies, and banded land snails (*Cepaea nemoralis*). In fact, he was working on snails when we started our collaboration on moths.

In addition to the many hundreds of *cognataria* in the collection donated to the Museum of Zoology, each specimen came with a label describing where it was caught, and *when*. It wasn't until my second year on the Reserve that I tracked down this collection. The Insect Division Collection Manager, Mark O'Brien, initially had trouble locating it in the vast museum holdings, because I requested it by using the current scientific name for the species, *Biston betularia cognataria*. Nope. Nothing by that name here. Well, as I pointed out elsewhere, taxonomic designations change as researchers learn more. So I tried a couple of the former appellations. Nope. Again, nothing here. Hmm? So then I switched to the common name: peppered moths (also called pepper-and-salt geometers). Yes! Here they are.

Drawer upon drawer of *Biston betularia cognataria*. No question about it. The Museum of Zoology had them, and Denis Owen had collected them, and 90% of them were black—fully melanic f. *swettaria*, indistinguishable from their black English cousins, f. *carbonaria*. I had never seen anything like it. To read a number, such as that 515 out of 576 were melanic, just doesn't have the impact of seeing hundreds upon hundreds of black moths right in front of your eyes. This was *real*, and I was moved by the experience. I also learned something else. The best time to trap *cognataria* on the George Reserve is May and August. Hardly any are flying during June and July.

In a fax sent to me by Denis the following year (on June 28, 1995), he wrote: "Your letter . . . arrived today . . . about the voltinism in *cognataria*. Yes, of course we have been stupid, and I should have known better because after all I collected the damned moths. Thank goodness I gave them to the Museum."

At the bottom of this letter, Denis added by hand Vladimir Nabokov's "disappointed lepidopterist's ditty":

It's a long climb
Up the rock face
At the wrong time
To the right place

43 | Edwin S. George Reserve

My 1992 Plymouth Voyager minivan was packed to its roof rack when I left my cozy home in Virginia on June 28, 1994, for my first visit to Michigan's E. S. George Reserve. I had several moth traps (both assembling and mercury vapor lamp designs); a gasoline-powered portable electric generator; hundreds of living moth pupae, held in dampened peat moss kept inside plastic shoeboxes; stacks of patterned cloth panels in both black and gray, from background rest-site test pens; screened mating cages; a computer and printer; clothes; bedding; and too much food. I was as prepared as I knew how to be. I couldn't have been more excited if I were Neil Armstrong heading for the moon. Perhaps I exaggerate. The moon was a much bigger deal, I admit. Neil Armstrong was a genuine hero of our entire species. That was also true for his colleague astronauts, valiant explorers every one! But in my small corner of this world, I was also an explorer of the unknown. While peppered moths themselves aren't dangerous, I almost crashed my overloaded minivan on a twisty mountain road in western Pennsylvania on my way to Ann Arbor, Michigan.

Once at the University of Michigan, I reported directly to the administrative secretary's office to pay fees, obtain keys, and get directions to the George Reserve. She gave me what I needed with barely a nod. It was as if she saw explorers like me every day. I never did meet the directors.

The 40-minute drive from Ann Arbor to Pinckney was quite pleasant, along a route with roadside farm stores that sold fresh fruits and vegetables. Just before arriving at the reserve itself, I passed though Hell. There is an oft-told story behind the name of this village in Michigan that I need not repeat, but spending time in Hell isn't nearly as awful as it sounds. I rather liked the place.

Almost seconds after leaving Hell, I arrived at the George Reserve, situated along Doyle Road in Livingston County. There was a modest administration building (a converted ranch-style house, painted flat green), an extensive garage and large workshop compound, an informal parking area, and the caretaker's splendid residence (a large white farmhouse). Just beyond the main entrance complex, the unpaved road into the reserve itself was secured by a tall, chain-link fence gate that would slide open only to authorized personnel. To become one of those privileged people, I presented myself to the caretaker, Jack Haynes—a very pleasant chap who was expecting me. He introduced me to the several staff members we encountered as he showed me how to work the security gate and gave me a brief tour of the grounds before taking me to my luxurious quarters (by field station standards): a small garage, converted into a furnished two-bedroom apartment, with a living room, kitchen, and bath. Not bad at all, and certainly more than enough room for one person. I had the place all to myself until Denis would arrive some weeks later.

The garage apartment was across a spacious lawn from "Hill & Dale," the main house on the reserve, which accommodated larger groups. When I arrived, the house was occupied by a team of graduate students and their mentor, Justin Congdon, a University of Georgia ecologist who was the director of a decades-long study of turtles on the reserve. We were introduced all around and got along very well. To paraphrase Will Rogers, I never met an ecologist I didn't like.

Now it was time for me to get to work. I unpacked my minivan and set up several moth traps using 100-foot extension cords, stretched in different directions from my garage apartment. I would soon spread out my operation to cover a larger area, but I didn't want to miss the opportunity to catch moths my first night on the George Reserve. Around midnight I heard the shrill scream of a raccoon that had touched its nose to a mercury vapor light bulb. The bulbs get painfully hot, and raccoons soon learned to avoid them.

The very first morning—the next to the last day of June—there were, among hundreds of other moths, three perfectly beautiful, typical *Biston betularia cognataria* in the traps right outside my apartment. I remember saying to myself, "This is going to be easy."

My optimism was fleeting. Easy it was not. Those three were the first and last peppered moths that showed up for the next four weeks.

After several nights of large catches of *other* moth species, I decided to try my luck elsewhere on the 1,300-acre George Reserve: the Hill & Dale front lawn; the Hill & Dale back lawn; the garage's back lawn; the field across the road from the garage; Building 13; by a pond below the barn; Evan's old field; both ends of Crane Pond; the north gatehouse; and the field near the south gate. Recalling the definition of insanity, attributed to Einstein, about doing the same thing over and over again with the expectation of different results, I had to at least vary the location of my traps from time to time to avoid going mad. Yet, no matter where I put my traps during the month of July, there would be multiple hundreds of moths of numerous species inside each trap, each night, but not a single *cognataria*. The groundskeepers wondered what I was complaining about: "Moths? They're all over the place!"

While I knew from Cyril Clarke's extensive records that June into July was the ideal time to catch British peppered moths on the wing, I would come to learn that it was the very worst time of the summer to find American peppered moths flying in Michigan. In frustration, I moved moth traps around the George Reserve like a crewmember rearranging deck chairs on the *Titanic*.

I first met Denis Owen face to face at the Detroit airport on August 1, 1994. By that time I had caught eight *cognataria*. I was more than a little embarrassed by that number, so I was unsure about how I might greet "Mister Parallel Evolution." I recognized him the moment he entered the reception area. Despite my neglecting to display a sign with his name on it, he identified me. We immediately approached each other with broad smiles. "It is easy to spot an ecologist in a crowd," he said approvingly.

As we shook hands, I replied, "I'm not an ecologist. I'm a geneticist."

"Well, you look like an ecologist," he said.

"Really? What does an ecologist look like?" I asked.

"Like you do," he answered.

"Well then, what does a geneticist look like?" I asked.

"Roger Milkman."

It's true that Roger Milkman was a renowned geneticist, but it's also true we didn't look alike. For one thing, Roger wore neckties.

That opening exchange set the style for our conversations over the next two years. We debated everything, whether silly or serious, and during the process we became good friends.

By the time we arrived at the George Reserve from the Detroit airport, Denis had decided I was an *ecological* geneticist. Even before unpacking his suitcase, we set about inspecting the moth traps I had set up for that night. Denis nodded his approval, despite knowing I had caught only eight so far. When we finished making the rounds, Denis brought out a bottle of Scotch. Dusk brought out mosquitoes. Denis was a smoker but refused to do so inside our garage residence, out of consideration for me. That first night we talked to each other through a screened window. Denis and the mosquitoes were outside. He was smoking. I was inside with his Scotch.

We patrolled the traps at several intervals each night, squinting to see what moths had entered. We paid no attention to the frenzied bats feeding on myriad bugs just above our heads—or to the mosquitoes feeding on us. Our focus was on moths. Denis would shield his eyes with his hand from the harsh direct light of the mercury vapor lamp while walking around the traps to get a feel for how the night's catch was shaping up. By morning we'd carefully unpack each trap to tally the results, but Denis was too excited to wait all night for that, so we'd take these periodic anticipatory tours. Denis was the most intense and knowledgeable naturalist I had ever met. He was the real deal.

I was also running assembling traps, using virgin females that had emerged from the lab-reared pupae I had brought along. In addition to this, Denis wanted to set up a white bedsheet stretched over a clothesline, illuminated by a mercury vapor lamp. This was the technique he had used on the reserve when he was a graduate student. It is also the method preferred by serious moth collectors, because the moths are carefully removed from the surface of the sheet as they come to rest on it and are put into killing jars (containing ethyl acetate). I didn't normally use this method for two reasons: (1) I wanted living moths, to assess their resting-background reflectance preferences, and (2) this method of trapping required the collector to stay up all night watching the sheet, in order to remove the desired moths when they alighted. I much preferred to use the Robinson method, which allowed me to run several traps at the same time at various locations. For collectors, who desire perfect specimens, the stretched bedsheet makes sense. For those of us doing population surveys, it doesn't matter if the specimens might lose a few scales inside a Robinson trap. We want numbers! In an attempt to collect a respectable quantity during the short time we would have on the George Reserve in 1994, we used every method of trapping we had available.

Each morning, after very little sleep the previous night from our rotating shifts at the stretched sheet, we'd make the rounds to unpack the several Robinson traps and the assembling trap. Despite our efforts, the daily yield was small. But we remained optimistic that we'd get an interpretable result, as we managed to catch at least a few peppered moths and hoped for more.

As I fumbled around our garage kitchen early one morning to get the coffee started, Denis greeted me with, "There he is, Mister *cognataria*!" That was indeed very generous of him. Was he passing me the baton? Then he said, "You'd better put a dram of whisky in your coffee this morning. There's a male in the assembling trap."

That night we celebrated. Denis enjoyed good food and wine, but he especially craved hamburgers made in American restaurants. McDonald's was not remotely what he meant. He wanted a hamburger from Hell.

So there we sat, at a small table in a dark room in Hell, drinking beer and waiting for our burgers. We chattered on, as we always did, about this and that. Denis generally bested me on most of our debates, partly because he was far more skillfully glib than I, but mostly because he simply knew more than I did about nearly everything. On matters of ecology or genetics we held our own in our respective fields, but on literature or the classics, Denis would kick my butt. On music, I kicked his. His excuse there was that he had too much to occupy him without having an unnecessary distraction, so he was willing to ignore music for the sake of his other interests. "But what about bird songs?" I asked. He conceded that was a significant deficiency, especially for an ornithologist, his primary affiliation. He allowed that he did pay close attention to bird songs, but not so much to Mozart. He in no way downplayed music, or the love of it. He just said it was a distraction he couldn't afford with all of his other interests. Part of this discussion developed from my playing the bagpipe on the George Reserve. He didn't disapprove. Instead, he was amused, insisting I must be the first and only bagpiper on the E. S. George Reserve.

Our hamburgers arrived. We avidly dived into them, with relish and kept on talking. At some point we got to the subject of species definitions and taxonomic assignments—starting first with North American pepper-and-salt geometers and British peppered moths. Although the species, *Biston betularia*, is actually Holarctic, its common nomenclature reflected regional chauvinism. From there we moved on to more-general questions.

I do not intend to relate esoteric details of our innumerable discussions here. I simply offer the following one as an example of how we went on, day and night! While we were drinking beer and enjoying hamburgers in Hell, Denis demanded to know why taxonomic relationships based on painstaking comparative anatomy should be revised by DNA sequences. I answered easily and quickly: "DNA analysis is more objective." He was stumped for a comeback. Rare! In that brief moment of silence, I noticed that people sitting at other tables in the restaurant in Hell were staring at us. Maybe it was Denis's British accent? Or perhaps because we were so loud?

44 | Farewell and Welcome

I took Denis back to the airport 10 days after he arrived. We had managed to improve our sample size (*N*) but still were disappointed we hadn't caught more. Denis's parting words were, "Fax *N*."

I stayed on at the George Reserve until I had to return to Williamsburg for the start of classes. My fax to Denis, sent August 24, 1994, not only included *N*, it also included this table, comparing his early records and our current data:

Years	*Typicals*	*Melanics*
1959–61	61	515
1994	21	4

$P \ll 0.001$

The difference in sample sizes for the years 1959–61 compared with 1994 is obvious. What might not be so obvious at first glance is the difference in the percentage of typical and melanic moths between the sampling periods. The *P* value in the table has nothing to do with the difference in the sizes of the samples. It is based solely on a comparison of the relative numbers of typicals and melanics between the two samples. By convention, a *P* value above 5% (or 0.05) would mean the differences in the distribution of categories (phenotypes) between the samples are not *statistically significant*—that is, the differences could reasonably be attributed to chance alone. In statistical language, a $P \ll 0.001$ is regarded as *very highly significant*—that is, the probability that chance alone accounts for the differences is very low. We therefore concluded with confidence that the frequency of the fully melanic f. *swettaria* had declined *significantly*.

If we compare the data from Denis Owen's last published sample at the George Reserve (taken in 1961), where $N = 24$, with our 1994 sample, where $N = 25$, the differences in the phenotypic distributions are equally whopping and perhaps easier to see, because of similar sample sizes:

Year	*Typicals*	*Melanics*	*% Melanics*
1961	2	22	92
1994	21	4	16

$P << 0.001$

The *P* values in both comparisons are the same. They are based on G-tests (under a null hypothesis, as in chi-square tests), but one needn't be a statistician to be struck by the huge drop in the percentage of melanic peppered moths at the George reserve in 33 years.

The decline in melanism we discovered was equally as dramatic as that observed by Cyril Clarke in England. In 1994, he trapped 348 peppered moths in his garden, of which only 18% were melanic. There was no significant difference between the percentages of melanics in the American and British samples.

These rapid changes happened over the same time interval, from effectively the same starting point (90% melanic moths), in populations separated by the Atlantic Ocean. Years before, Denis Owen had documented the parallel *rise* in melanism in both countries. Now we had just discovered its parallel *fall*. This was *two-way parallel evolution*—a rise and fall of melanism on separate continents—involving phenotypes caused by alleles at the same gene locus in geographically isolated subspecies. Was there another *known* example like it on this planet, we wondered?

All three of us recognized that our discovery was huge! As collaborators, we had to decide where and when to publish. Our sample of 25 moths was small, so we agreed to return to the George Reserve to get more data. Of course, every study could always use more data. My colleague at William & Mary, Charlotte Mangum, dismissed as "too boring" those studies designed solely to gather more data when existing data were unambiguous. Nonetheless, Denis, Cyril, and I resolved to do just that. But obtaining a bigger sample couldn't happen until the moths were flying again. Should we sit on what we had and wait another year? Denis was disinclined to do that. Cyril was feeling his age and said emphatically, "Maybe you have lots of time ahead of you, but I don't." So we agreed to announce our discovery and justified our moving so fast as perhaps motivating other moth collectors to get involved. Why waste another year?

The first journal that pops into scientists' heads when they think they have something important to say is *Nature*. If not *Nature*, then *Science* will do. These rank as the number one and number two publications in which to announce world-shaking scientific discoveries. Because everyone in the business knows this, they all try to publish in *Nature* or *Science*. Getting a manuscript accepted in either is a feather in one's cap, because competition is so horrendously stiff, especially for young scientists seeking jobs or tenure. Prestigious journals obviously can't accept all submitted papers. They reject most of them, often without even bothering to send them out for review. Why should they squander a reviewer's precious time with a manuscript that is of little interest to the journal's readership? It's a marketing decision.

Nevertheless, Cyril pushed hard for *Nature*. Denis agreed, pointing out that his field-defining article, "Parallel evolution in European and North American populations of a geometrid moth," was published in *Nature* in 1962. "Surely they'd welcome a follow-up report about that."

I don't recall the exact date when we submitted our manuscript to *Nature*, but it was before a letter to me from Denis (dated September 16, 1994) probed, "I hope you sent off the paper." *Nature* acknowledged its receipt straight away. After that we heard nothing for a long time. Maybe no news is good news, meaning our manuscript passed the first hurdle and they'd sent it out for review? Denis said in his letter, "I think the paper stands a good chance with *Nature*."

At first, not hearing did seem to us like a good sign. I still hadn't found out anything by the time Denis and his wife Clare came to Williamsburg in October. As I greeted them at the Norfolk airport, Denis asked, "Still no word from *Nature*?"

Their visit was more social than professional. We hadn't known each other for very long, but something between us clicked. We both enjoyed a debate, and we each saw a challenging opponent in the other. But mostly we had fun. Adding Cathy and Clare to the mix made our interactions even more enjoyable. We all became better acquainted and relaxed in each other's company.

On their very first night as our houseguests, Denis asked, "Do you have a moth trap?"

"Of course," I answered. "I don't usually run it this time of year. I've never caught *cognataria* here in Williamsburg, and their season is over in this part of the world."

"I wasn't actually thinking of *cognataria*," said Denis. "Surely there are other interesting species about?"

That settled it. Into the garage we went, to get my best Robinson trap. We set it up on the deck before we sat down to supper. As the sun set, we were bathed in brilliant light from the mercury vapor lamp. Denis sighed, "Ah, now that's better."

Denis presented a seminar to William & Mary's Biology Department about his extensive ecological work. He illustrated his talk with examples from an extraordinary variety of organisms, with barely a mention of peppered moths. He was a crowd pleaser and took the opportunity to get in a few jabs at me, much to the delight of my students and colleagues in the audience. But back in my office, we focused on one thing. We worked hard at digging into the literature for the merest hint of anything related to our research collaboration, and we made plans for the next year. We needed hard data on atmospheric changes in the Michigan region to compare with the British studies. We learned of additional data about melanism in American peppered moths in Pennsylvania and New England, and we established contacts with American lepidopterists. We designed our work schedule for the next year.

Yet we were uneasy, because we still hadn't heard anything from *Nature*. What's going on with them, we wondered? Denis suggested that we should redraft the paper for another journal and mentioned several he thought were appropriate. I agreed that a backup plan would be wise, but I was not ready to throw in the towel. He said, "You certainly are determined!" I took it as a compliment.

We also spent some recreational time having a look around southeastern Virginia. Clare loved to collect seeds of all sorts to take home and plant in her garden in Oxfordshire. I teased that she was perpetuating the ecologically disastrous British tradition of introducing exotic species from all over the world to all over the British Empire.

We had a day trip to Hog Island to check out the bird colonies there, as the fall migrations were just starting. Clare was an avid bird watcher. They first met when Denis, who was a professional ornithologist, led a tour she joined as an interested amateur. I would later learn that they were newlyweds, married the same year we began our moth collaboration. Both thoroughly enjoyed birding and were much better at it than I was. On one of our local excursions, we spotted a mixed flock of songbirds that suddenly flew off in all directions, as if startled. I asked, "Wow, what caused that explosion?" "Probably that Cooper's hawk coming in from the left," Denis replied. Indeed, there it was. A menacing raptor had entered the scene. I hadn't noticed it until Denis pointed it out.

He saw stuff like that. All good birders do. I tip my hat to them. Yet Denis was far more than a birder. He was a disciplined observer of nature, and he helped me better appreciate field biology. Before I became so involved with peppered moths, I thought field biologists wore flannel shirts, ate granola, and slept in the woods. Well, that part's still true.

We also went to Virginia Beach, but not to swim. Denis wanted to look for *Donax variabilis*, a widely distributed, small saltwater clam known for the highly variable color patterning of its valves (shell)—hence the species name. He presented geographic distinctions as evidence that reflexive polymorphisms are maintained by visual selection, because birds are unable to form search images for prey that differs discontinuously from nearly every other member of its population. His interpretation and choice of terminology did not go unchallenged, as others preferred "apostatic" (negative frequency–dependent) selection as the mechanism—that is, the advantage of being rare. It was a quarrel I avoided, although the clams themselves were amazing in their variability.

Within a few minutes Denis had excavated several meters of the beach and had hundreds of clams in plastic buckets. A half dozen or so fit easily across the palm of his hand. No two that I saw looked alike, although they could be grouped into broad categories of hues, shades, stripes, etc. While Denis was standing in one of his sand pits, surrounded by numerous others he'd scraped open, two women (with children in tow) came along. One of the ladies regarded Denis curiously and demanded, "What are you looking for?"

"*Donax*," Denis replied.

"Don't ask?" the woman repeated loudly. "Why not?"

The other woman inquired, "What'd he say?"

"He said, 'Don't ask'! Can you imagine that? Well, *excuse* me!"

Denis thought this was all great fun, although he did show them the tiny clams in his hand. He then explained the scientific name for these clams, added a few of their common monikers, and gave a brief lecture about how natural selection works. The women were very receptive to his informative talk and kept saying, "Hmm. Now that *is* interesting." Their kids were fascinated by the clams and dug up quite a few more for Denis. Then one asked, "Are they good to eat?"

"Indeed they are, but it would take quite a lot of them to make a meal. Are you hungry enough to dig for your supper?" Denis was a natural teacher, and he enjoyed that role enormously.

The women thanked Denis, and as they moved on, I could hear them agreeing about how informative it all was. One said, "At first I thought he was a bum."

45 | *Nature*

It wasn't until February 16, 1995, that our scientific correspondence announcing two-way parallel evolution in peppered moths appeared in *Nature*, under the unassuming title "Decline of melanic moths." We got what we wanted, but it was a bumpy ride. All published authors in refereed journals have complaints about demands by reviewers. None proved fatal to our manuscript, but we were made to shorten it. Fair enough. We only had an *N* of 25, so we were pleased that *Nature* appreciated how our 25 differed from the 24 the last time the population was sampled, 33 years ago. Besides, James Watson and Francis Crick also got only a single page in that journal in 1953 to present their discovery of the structure of DNA. We were in good company in that respect.

What really held up the publication of our paper was that *Nature* mislaid our file. All of this got sorted out eventually, so there is not much point in detailing the relevant correspondence. Far more interesting is the reaction to the note by our colleagues. Several questioned our suggestion "that the rise and subsequent fall in melanism in the *Biston* population of the George Reserve is not related to lichen succession" (p. 565). We based that statement on Denis Owen's personal observations (his "before" and "after" recollections of the appearance of the habitat) and—far more authoritatively—on the personal communication we solicited from Howard A. Crum, professor of botany at the University of Michigan and a world-renowned bryologist, who taught courses in bryology and lichenology. He reported there had been no perceptible change in the lichen flora in that region of Michigan over the relevant 30-year span. Still, several prominent experts in England, who had never even been to the George Reserve, insisted we'd got that wrong.

One especially well-respected colleague had written immediately to Cyril to express his doubts. Denis replied on March 20, 1995:

> Cyril Clarke has given me a copy of your letter (27 February) and suggested I might respond.
>
> It is true we are unconvinced by lichens affecting *Biston*, and we are not at all sure about background. *Nature* wouldn't give us enough space to elaborate, and they are probably right, as we really need more information.
>
> So, we intend to sample the Michigan sites twice this year (the moth is double brooded there—June and August) and after that we'll try a more substantial paper.

The letter went on to ask about other matters, but the die was cast. We were going back to Michigan again. The only thing Denis had got wrong is that waiting until June is too late for the first brood of the season. It would have been much better if we had gone back in May. We would learn that when I examined Denis's collection after I went back to Michigan in June for round two.

46 | Round Two

Back in my Plymouth minivan for a second attempt, I left for Michigan a month earlier than the previous year. I took even more equipment in 1995, adding an extra Robinson trap borrowed from Lincoln Brower. The trap originally was a gift to Lincoln from Denis, but Lincoln wasn't using it and happily donated it to our project. I also brought along some Tris-buffered saline solution, to store freeze-killed moths in vials for later DNA analysis. I had an undergraduate research student, Jennifer Sekula, who hoped to examine mitochondrial DNA (mtDNA) to assess genetic divergence between American and British peppered moths. Her lab work was under the supervision of John Graves, an established expert on mtDNA at VIMS (William & Mary's School of Marine Science). Denis, using the same saline preparation, was to provide moths for the British DNA samples.

We clearly had bigger plans for the summer of 1995 than we'd had for 1994. We also arranged to obtain atmospheric records for SO_2 and for suspended particulates (an indicator of soot accumulation), to compare them with British records over the appropriate years and locations. And we would interview other moth collectors in the region. We hoped to leave no stones unturned.

Once back on the reserve, I again had the garage apartment across from Hill & Dale house. The same crew from the turtle project was back and teased me with their greeting: "Look at you! Only one summer here and you've got a paper in *Nature*! We've been here for years and what do we have to show for it?"

"Not a fair comparison," I protested. "The work I took part in last summer was actually started in 1959. When did you start?"

It was all good-natured ribbing, but they did seem to show more interest in what I was doing. So did the Museum of Zoology, when their promotional blurb about research opportunities at the E. S. George Reserve featured a description of our ongoing moth work. Being a celebrity felt good.

The next morning felt even better. On my first day, June 1, I caught two *cognataria*, one typical and one melanic. Unlike the previous year, I didn't say to myself, "This is going to be easy." I had no such confidence going into round two. But I slowly became more optimistic as our *N* slowly grew. It didn't grow by much, however, when the first brood of summer finished flying by the middle of June. I would return in time to hit the ground running for the start of the second brood in late July. Denis would join me later.

Denis had been running moths traps in England and shipped parcels of Tris-buffered *Biston* specimens to me by air express. As he wouldn't arrive in the United States until August 6, and my research student Jennifer was to leave for law school on August 17, hand-carrying the sample wouldn't allow her nearly enough time to complete her work. Denis mailed his last shipment on July 15, followed by a fax ending with, "Looking forward to the hamburgers in Hell." Clearly Denis was excited about being back on the George Reserve for "another go." It had been on his mind even before our paper came out in *Nature*. In a fax sent on January 12, 1995, he said: "What do you think about Michigan in 1995? It would be fantastic to go twice, once for the first brood and once for the second, but this may present you with problems. I might get money for at least one trip, but thinking about it I am completely dependent on your being there, too (traps and everything else). Possibly if we do one trip it could be in August: what do you think?"

Our second meeting at the Detroit airport was much different from the first. No period of adjustment required. By round two, we knew each other well. Besides being professional colleagues, we had become good friends and got along very well. Even our disagreements, all of which were very minor, were settled easily and with humor.

On the drive back to the reserve from the airport, Denis happily announced that we each had been awarded travel grants of £1,000 by the Nuffield Foundation, arranged through Cyril Clarke's grant. Nuffield had supported Cyril's work for many years, and he recognized that his role in this project cost him virtually nothing, while Denis and I had considerable out-of-pocket travel expenses. Every little bit helps, so I exclaimed, "Great!" Denis went on to explain that to make matters easier for me, he converted my £1,000 into American money, although there would be a small subtraction, due to exchange rates and commissions, to which I said, "Understood." Then Denis went on to suggest that I might use some of my new travel funds to come to England in the fall, so we might work on our paper. The idea of my going to England in the fall was not new, but that there would be funds available sweetened the pot. It then occurred to Denis that his converting the £1,000 into American dollars might have been premature, because I would require British currency once I was back in the United Kingdom. I'd have to exchange American dollars to get British pounds sterling, and that would require again paying the exchange rates and commissions. If Denis had simply handed me £1,000, I'd have lost *zero* from the total instead of paying exchange rates and commissions *twice*.

Denis apologized and admitted, "I wasn't thinking."

My first impulse was to say, "Agreed." And to that I'd have added, "At least we agree on some things." But I tamped down the urge and said instead, "You meant well." That was true. After a few moments, I could resist no longer and said, "Good thing we're not business professors."

We resumed the work pattern we had established the previous year but increased our effort to build up the sample size by running a total of eight traps each night (the stretched sheet, four mercury vapor lamp and two black-light Robinson traps, and an assembling trap).

In addition to riveting our nighttime attention on moth trapping, we made several significant day trips. Of particular importance was an interview with Ms. Deborah Harren, to request annual reports issued by the Air Monitoring Unit of the State of Michigan's Department of Natural Resources for data about atmospheric SO_2 and suspended particulates, measured by sampling stations nearest the George Reserve that had operated continuously, from the present back to as long ago as past records were available. Ms. Harren came through for us, but it would take a while for her to gather that information, as a number of the records from single locations were fragmentary, since sampling stations were periodically relocated. The sampling station closest to the reserve for *continuous* SO_2 data was in Detroit (to the east), and for suspended particulates, Lansing (to the northwest). The George Reserve is about midway between the two, circa 45 miles.

Another of our day trips was to Michigan State University in East Lansing, to visit Mogens "Mo" Nielsen. He was Michigan's most well-known lepidopterist and proved to be a great help to us, both from his personal field experience and through his contacts with other collectors in the state. Denis and I spent a very pleasant day with Mo and felt much enriched by the experience.

Some of our days on the reserve, waiting for nightfall when the excitement would begin anew, were spent exploring the grounds. Just for fun, Denis found a long-handled net and walked across a field, sweeping the net from side to side in the tall grass. He then stopped to examine his catch. It was brimming with all sorts of critters, and Denis could tell interesting stories about so many of them. I'd seen most before, but I knew little about them.

One afternoon, the turtle crew brought back an enormous snapper. Denis lifted it, to feel its heft. The turtle extended its neck to get a better look, first at Denis, and then at the rest of us who were staring at it. One young woman mused, "I wonder what's going through its mind?"

"Whatever it is," Denis offered, "it's going through very slowly."

Lazy days, busy nights, and occasional hamburgers in Hell would best describe our activities, with intermittent side trips. On one occasion, we were in a lab building on the reserve. Denis had found a small field library and uncovered the thesis written by his first wife, Jennifer. He read it, and it brought back memories of his time on the reserve as a young man, a budding scientist with his long, productive career still mostly in front of him.

Some of these recollections made Denis a bit melancholy. When we needed to head to the airport for his return flight to England, I went searching for him. I found him sitting quietly alone in the Hill & Dale house. By this time the turtle crew had packed up and gone, so Denis had taken the opportunity to look in on his former year-round residence as a graduate student. "Time to head out," I said.

"Right," he answered, as he slowly pushed himself up from the armchair. "It's an odd feeling to know this is the last time I'll be here."

I caught three *cognataria* typicals on August 19. No additional *cognataria* showed up for the remainder of the month. From that, coupled with the information from Denis's detailed early records, it was obvious the season was over. I left with no regrets and a 1995 $N = 35$. More importantly, the phenotypic distributions between 1994 and 1995 confirmed that the previous year's result was not a fluke. The decline in melanism at the George Reserve was clearly real. The combined percentage of melanics (18.33%) was virtually identical to that at Caldy Common in England (18.23%) over the same two-year period (1994 and 1995). The parallel rise and fall in melanism in America and Britain was corroborated by repeated sampling.

47 | Oxfordshire

I slept late my first day as a houseguest of Denis and Clare Owen. By the time I awoke, Denis had already gone to work, and Clare had police in her kitchen. Clare had called the cops when she discovered her car had been broken into and vandalized during the night. From the damage, it appeared as though someone had tried to remove the car's stereo and, perhaps frustrated by failure, destroyed the instrument panel. The police offered little hope the culprits would be caught. Such crimes were all too common and left few trails. Clare would need a loaner car while hers was being repaired.

When I lived in the Merseyside area a decade earlier, burglaries were commonplace. The natal area of the Beatles was also the home of high unemployment and widespread drug addiction. Everyone I knew had multiple locks on their doors. It was an all-too-familiar part of urban life. But what happened to Clare's car shocked me. When I arrived at their home on Little Wittenham Road on the outskirts of Long Wittenham, a rural hamlet in Oxfordshire, I was immediately struck by its charm. Many of the houses had thatched roofs! I half expected hobbits to come out of some of them. It was as if this locale had escaped the passage of time. How could Clare's car be maliciously vandalized in a place like this? Clare kept calm and carried on.

Following that episode, Denis parked his car tight against the tall, thick, boxwood hedgerow separating their garden from the road, the idea being that to break into his vehicle, intruders would have to enter from the road side, thus risking being observed by passing motorists. Denis was determined not to allow any hiding space between his car and the hedge, as a deterrent. Sometimes he parked so close to the hedge that when he closed the vehicle's electric windows at night, some of the branches would get caught inside. When he'd leave the next morning, he'd rip off a clump of greenery and have it hanging from the side of his car on his way to work.

I would later get a car of my own, so Denis wouldn't have to chauffer me around, and I followed his advice about parking tight up against the hedge.

It worked! No one vandalized my vehicle while I was in England. At least, not that I could tell.

The primary purpose for my visit was to write a paper with my research collaborators, and I wanted to get started without delay. The plan was that I would write the initial draft, and they would read it over and make suggestions. That's how we coauthored the *Nature* article, but it was far shorter and much less complicated than our plans for a fuller treatment, so we felt a need for more direct contact. Besides, I had arranged a sabbatical to spend the fall semester as a visitor in Denis's department. I made no demands on Oxford Brookes University, nor felt any need for office space there. I simply wanted the use of a computer and printer to write the paper. As both Denis and Clare would be gone all day, I thought their generous offer that I use their home office was ideal. Denis proudly showed me the tidy workspace, including its manual typewriter. "It's all yours!"

It would do nicely, but I asked, "Where's the computer?"

"Oh? We have no computer at home. I thought you were bringing one along," he said, somewhat puzzled.

I had brought along a laptop, quite primitive by today's standards, and a portable printer. As I had already completed the regression analyses on the atmospheric data, created accompanying graphs, and run programs to find the curves generated from constant fitness models that best approximated the observed declines in melanism, all I really needed at that point was a word-processing program. My laptop had that. I would make do. Or so I thought.

A problem erupted when I plugged my equipment into a voltage transformer. Visible sparks shot out from the connection box, accompanied by an ominous hum. Then it went dead! Fortunately, the transformer had blown a fuse, so no damage was done to my gear. I had, in fact, used this very transformer, without incident, on previous trips to the United Kingdom. I could easily get another fuse, but I was just a little gun-shy about having another try at using this transformer, not knowing why the fuse blew. I strongly suspected a short, but where?

Clare took matters in hand. She borrowed a surplus (i.e., retired) PC from her lab for my use. She worked in the Department of Archaeology at the University of Oxford, and her lab specialized in radiocarbon dating.

Only a few years prior to my visit, her lab showed that the Shroud of Turin merely dated back to medieval times, far too recent to have been the burial shroud of Christ. Clare took pride in that work. Perhaps this borrowed computer played a part in demoting the Shroud of Turin? And now it would have a role in publicizing parallel evolution! God works in mysterious ways.

I'm not sure I had even written a paragraph before the weekend arrived, and Denis and Clare began taking me on outings to see the sights. The most memorable was our visit to Down House, Charles Darwin's home in Kent. While Denis and I might argue about whether the sun rose in the east and set in the west, the one thing we totally agreed on was that Charles Darwin was the most important person who ever lived. Period!

We had a tour of the house, its garden, and the Sandwalk, the path Darwin strolled daily while thinking. We stood in reverence in Darwin's study. We knew, of course, that the house and its furnishings had been restored to resemble their original appearance as accurately as possible, almost as if Darwin had just left for lunch. Denis said, "Darwin wrote the *Origin of Species* in this very room." I didn't feel the need to say anything. Denis continued, "And he didn't have a computer." Then added, "He didn't even have a typewriter."

Luckily for me, I had both a typewriter and a computer, and my task was minuscule, compared with Darwin's. He'd done all the heavy lifting. What was different about industrial melanism, in contrast to other examples of natural selection, is that the changes that occurred in peppered moth populations were observed by people *as they happened*, and the reasons were fairly well understood. What we had to offer that was new to this familiar story was a *replicate* example from nature: parallel evolution in *two* directions, on separate continents. All we had to do was write up our observations and publish them. For that task I had a computer.

By day I pecked away. When Denis came home from work, we'd discuss my progress before dinner. Clare was a superb cook. She had run a B & B in Greece some years before and was truly a pro. Denis's brother was a wine importer, and Denis kept a well-stocked cellar. Mealtimes were extraordinary, and they were always lively with conversation. Sometimes we'd simply go to the local pub, a short walk from the house. There Denis would insist I try the various ales available only by draught, and only in certain places and times of the year. I had no say in what beer we might choose there. I was permitted to pay, but not to order. Occasionally I went to the pub by myself, just to be alone, and to give Denis and Clare some free time. On one such occasion the bartender greeted me with: "So, they let you out on your own for the evening? What'll you'll have?"

"I have no idea," I said. "Do you have any Budweezer?"

Denis took me to meet his colleague Derek Whiteley, an accomplished artist employed by his department to produce technically precise illustrations for publications. The reason hand-drawn artwork is preferred to a photograph is not nostalgia for tradition, but because it allows a skilled artist to bring into focus all of the relevant details necessary for the identification of species, something rarely possible in a single photo. It is a very special talent, which Derek abundantly had. We went to his home studio. He was working on a side-by-side comparison of American and British subspecies of *Biston betularia*, which showed the typical forms (generally the American version is darker than the British) at the top of a panel, the two melanic forms (which are indistinguishable) at the bottom, and examples of intermediates (*insularia*) in the middle.

As mentioned earlier, *insularia* is a name applied to a range of intermediates—the products of multiple alleles—that some specialists graded, as did Denis. Cyril preferred to lump them. For him, a moth was either a typical or a full melanic; if not, it was *insularia*. For us it was not an argument worth having. *Insularia* forms, of whatever color gradation, were not common at the locations we were studying. Effectively we were dealing with two qualitatively distinguishable phenotypes: typicals (the lightest pale form for its part of the world) and full melanics. Any misclassifications of *insularia* would be rare, as well as of little consequence in our analyses. Elsewhere in Britain, *insularia* forms were more common, and Laurence Cook and I would later collaborate in their analysis in a paper we coauthored for *Heredity* in 2000.

Derek's artwork was exquisite! Denis commissioned it, to be included as a plate in our paper, illustrating the moth phenotypes within and between populations. "It's much too good for that," I exclaimed. "This belongs on a cover!"

With a draft of our paper—text, tables, and figures, including Derek's art-work—Denis and I went to visit Cyril and Lady Clarke at their home in West Kirby. We also arranged to meet Mark Seaward there, to examine the then current state of lichen succession along Caldy Common. Not much came from that meeting, but we were very grateful to Mark for his generous donation of time and expertise as a lichenologist. Cyril thought the manuscript draft was fine, and said "Yes" from time to time as he turned the pages. Lady Clarke, an artist herself, made suggestions to improve Derek's representation of British typicals: "They are too dark. I wouldn't have recognized them without the caption." Other than her husband, probably no person in history has seen more peppered moths than she had. Theirs was the largest and longest continuous record ever made, and she was involved with it every step of the way. We would most seriously carry her message back to Derek and ask him to lighten the appearance of the British typical in his drawing.

Not many journals in 1995 had illustrated covers. Of those appealing to the audience we sought—*Evolution*, *American Naturalist*, *Genetics*, *Heredity*, and *Journal of Heredity*—only the *Journal of Heredity* presented photos on its cover. Even there, we were uncertain that hand drawings in the exacting style of taxonomic comparisons of insects would break through that journal's long tradition of displaying actual *photographs* of organisms for its cover story. We hoped the editors might make an exception for Derek's work, because it accompanied our unique example of parallel evolution in two directions. We hoped they'd see it as a big story.

After we had submitted our paper to the *Journal of Heredity*, Denis and I traveled to London by train to attend a Natural History Society gathering of moth collectors. Our mission was to update the survey of the incidence of melanism in peppered moth populations throughout the United Kingdom and, as far as possible, the United States. The last national survey in the United Kingdom had been done over a decade previously, through a program involving students from the Open University, who were asked to collect moths over five-day periods. Unfortunately, sample sizes averaged fewer than 10 moths per sample site. Still, the organizers of the survey, Laurence Cook et al., reported declines in melanic frequencies at some localities. We thought it was time to have another look, and we hoped to recruit experienced collectors to the cause by making an appeal at the London moth meeting. Denis said, "He who ignores amateurs does so at his peril." He pointed out that the great success of Kettlewell, himself an amateur, was to enlist the services of moth collectors nationwide. Denis hoped to reach some of these same people, as well as attract new ones.

I had never been to a meeting quite like this. It was a show-and-tell exhibition, held in several large rooms, with rows of long tables displaying pinned specimens—the prize catches of numerous collectors. It was dazzling to behold.

By each display, the proud owner was on hand to explain what he or she had to show. These people were extremely knowledgeable about their specimens, as well as about other species, where and when to catch them, and by what method. Many of them either knew Denis personally or by reputation. And they welcomed me as his colleague.

Denis had the only display of pinned peppered moths. Compared with the other exhibits, his was rather colorless. Nevertheless, Denis could spread wings and pin moths as well as anyone, and he choose gorgeous specimens of both American and British *Biston betularia* to demonstrate, side by side, the polytypism and polymorphism in their wing patterns. In fact, these were the very specimens that served as the models for Derek Whiteley's drawings. Using a large typeface, Denis included in the display the title of our paper, "Parallel rise and fall of melanic peppered moths in America and Britain," along with an excerpt from the abstract. It got attention.

Denis and I took turns being stationed by our display to buttonhole passersby, but the whole affair was very informal, so we also milled about, looking at other displays and drifting from conversation to conversation. It was a pleasant gathering of people with common interests. They even had a place on site where one could get sandwiches, snacks, and beer. Two friendly chaps invited me for a pint. I ordered a lager. I had barely taken two sips when Denis happened by the bar and saw what I was drinking. He shook his head in disapproval at my disloyalty, especially after all he'd taught me about proper British ale. "Bruce, Bruce, Bruce," he said, "I leave you on your own for two minutes and look what you do! Lager? Is that what you've got there?"

At the end of the day, Denis called out in a loud voice to get the attention of the moth assembly before they headed for the doors. He explained his need for their help to conduct a national survey of melanism in peppered moths. He asked that they take notes of their catches and send him their data at the end of the season. For any who might have doubts about the scoring, please just send him the specimens. He finished by saying, "Let's make 1996 the year of the peppered moth!"

48 | New York Times

Unfortunately, 1996 was also the year Denis Owen died. His passing was announced in America by the *New York Times* on November 12: "Dr. Owen, who died last month . . . " That bit of information was included in the featured article—the top story—of the Science Times section for that week. The artwork was spectacular, taking up most of the front page of that part of the newspaper. It was very cleverly adapted from Derek Whiteley's illustration for the cover of the *Journal of Heredity*, and they acknowledged his work as the source. "PARALLEL PLOTS IN CLASSIC OF EVOLUTION" was the headline for the article written by Carol Kaesuk Yoon, who opened with: "It is probably the most famous story of evolution observed." She got that right.

In fact, she got just about all of it right. I was highly pleased with the coverage. She quoted me throughout, and my mother was very proud of that. As one of the coauthors, naturally Cyril Clarke was also cited. And the article included remarks from several prominent evolutionary biologists who were not directly involved with the research. Most good news stories about scientific developments seek outside opinions from experts, and the *New York Times* leads the pack. We were pleased that John Endler, then at the University of California, Santa Barbara, was quoted as saying, "It's an extremely interesting study," adding that it was the only example he, like other researchers, knew that documented such a parallel reversal—the same evolutionary about-face in two very distant locales. Douglas Futuyma, at Stony Brook University, said that the new study served to strengthen the classic story: "When you have parallel changes like this, it's like having different replicates in an experiment. The more you have, the more confident you are that you're getting a consistent result." And, from Richard Harrison at Cornell University: "Other cases of natural selection just don't jump out at most people."

All of that was well and good, and Denis's prominent role from the very beginning of the American side of the story was related beautifully by Ms.

Yoon. She did an excellent job of reporting. The only problem was that Denis never got to see a word of it. He died on October 3, just over a month before the *New York Times* recognized his work, which was begun as a graduate student, and only four days before the printed version of our paper came out in the *Journal of Heredity*.

49 | Expanding Views

The 1996 season for *Biston betularia* was well over by the time Denis died in October, although he continued to run moth traps in his garden to within days before the end of his life.

About a year earlier, he had developed a persistent cough that was attributed to bronchitis, exacerbated by heavy smoking. On doctors' orders he quit cold turkey and popped Polo mints to suppress the urge to smoke. He was successful, but it wasn't easy. Still, his cough got worse. Within two weeks of the diagnosis that his bronchitis was actually lung cancer, he was dead. The official cause was the pneumonia he contracted while undergoing treatment for cancer. His immune system was weakened by the procedure, and the pneumococcus that attacked him was resistant to antibiotics—a modern superbug produced by natural selection. Although lung cancer was a fairly certain death sentence, the suddenness of Denis's demise shocked us all.

By that time, Denis not only had his own peppered moth data for the 1996 season, he had also received data from a number of people he had recruited at the moth meeting we attended the previous year, as well as through other appeals he'd made in entomological newsletters and bulletins. The data were coming in.

The task fell to me to analyze and publish it all. Clare sent me what information Denis had at the time of his death, as well as whatever other reports she received. Denis's brother, the wine merchant, was also an amateur moth collector, and he asked other collectors to send their data to me. In all, I received the 1996 collection records from 40 moth workers throughout the United Kingdom. The nationwide survey Denis had organized would go forward as planned.

The sample sizes ranged from 3 moths to 373. From these I selected only those reports with sample sizes greater than 25, making one exception to this minimum cutoff in order to balance geographic representation. In all, 18 sampling locations composed our 1996 national survey, for a total of 1,642 moths, with an average sample size of 91 moths per location.

The point of this survey was not simply to assess the current geographic variation in melanic frequencies across Britain, but also to determine if the precipitous decline in melanic frequency that Cyril Clarke had documented at Caldy Common was a widespread phenomenon, or if it was restricted to the Liverpool area. To answer that question, I compared the 1996 melanic frequencies at our 18 locations with those at the most closely matching locations reported 40 years previously by Kettlewell. These paired statistical comparisons were not exact matches, but they were as close on the map as we could manage.

For visual impact, I abridged Kettlewell's map of Britain, with pie charts indicating melanic frequencies at these locations, to present a "before" image alongside an "after" map with pie charts for 1996. It was a really striking comparison, which made it clear at a glance that the decline in melanic frequency happened everywhere in Britain. While there were still obvious geographic variations across the country, in *every single place*, melanism had gone down, except where it was absent to begin with (e.g., in Dorset and northern Scotland). As the differences in the "before" and "after" distributions were statistically highly significant, we concluded with confidence that the decline in melanism in British peppered moth populations was widespread—and real.

The national survey paper we published in 1998 had four authors, including Denis. Since that time, journals no longer allow including deceased persons as authors (unless the paper had already gone to press). Denis died before we even gathered the data for this paper, but we felt strongly that he should be listed as an author, because he was the one who organized the British survey. The convention today would be to identify his contribution in the acknowledgments. Understandable. Cyril Clarke was also listed as an author, of course, because his Caldy Common data anchored the study, and he read and approved drafts of the paper. A new coauthor joining us, A. D. Cook, is not to be confused with the renowned Laurence Cook. This Cook was then my undergraduate research student, Amanda, who has since become a forensic psychiatrist. Amanda's role in this project was to trap *cognataria* in southwestern Virginia and eastern West Virginia.

Unlike Denis in the United Kingdom, and Kettlewell before him, I did not have (or know of) a network of amateur moth collectors to call upon to assist with geographic surveys in North America. At whatever location where I needed samples, I had to go there and run moth traps to get them. The one exception was David West, who ran traps at the Mountain Lake Biological Station and at his home in Blacksburg. When Amanda volunteered to run traps where she lived, in Tazewell, Virginia, and in the surrounding area during her summer break, I happily accepted her offer. While she was there, I ran traps in Pennsylvania.

T. R. Manley had recorded melanic *cognataria* near Klingerstown in Schuylkill County, Pennsylvania, between 1971 and 1986, during which time melanism fell from over 50% to 38%. The county, though rural, is strewn with coal strip mines. We were eager to trap in that area, to see if the decline in melanism there had continued and compare it with data for southeastern Michigan. The location proved to be convenient for me, because I had access to my parents' home in the anthracite coal region near Wilkes-Barre and my in-laws' farm in rural Columbia County. Ultimately, I would trap the entire area, including Schuylkill County. But for the 1998 paper, our American sample locations were few and far between. Nevertheless, we saw essentially the same pattern of decline that had occurred in Britain. Where melanics were once common, they had become significantly less so, and where they were rare, they remained infrequent or disappeared altogether. The atmospheric records provided by Pennsylvania and Virginia governmental agencies showed a decrease in SO_2 in Pennsylvania, in concert with the decline in melanism there. The assessments of SO_2 in Virginia stayed about the same, but it was always significantly lower than Pennsylvania's levels. We focused particularly on SO_2 in this paper, because Mark Seaward had shown, through a multiple regression analysis, that it is the most important of 13 environmental variables related to melanic moth frequencies. We were not attempting to identify the causative agent for the decline, but we did show that what appeared to be happening in American peppered moth populations was consistent with what was occurring in the United Kingdom, and it was similarly correlated with changes in atmospheric SO_2.

Despite our best efforts, there remains to this day a general unawareness—even among professional evolutionary biologists—that peppered moths exist in North America, and that they have a history of melanism. True, far more attention has been paid to British peppered moths than in all of the other places combined where this Holarctic species lives. By comparison, the American story is woefully wanting in details. In the hope of filling in some of the blanks, I recruited my colleague in the biology department, Larry Wiseman, to lend a hand.

Larry is a developmental biologist, whose research at that point focused on how cells sort themselves out in animal embryos. Many modern biology departments group people like him in their molecular/cellular subdivision, and people like me with ecologists and evolutionists. Fortunately, at William & Mary (at least when I was there), we saw biology as a multifaceted discipline and insisted our students get the whole picture. That required a faculty at least conversant with the broad range of fields across the whole of biology. Certainly for research, and for teaching assignments, we were also specialists. This is as it should be, and it is not uncommon. James Watson, for example, was a bird watcher as well as a molecular biologist. At Mountain Lake, I occasionally used Dietrich Bodenstein's moth trap. He was a developmental biologist of great renown. Yet his hobby was collecting moths. Sorry I missed meeting him! The point I'm hoping to make is that Larry Wiseman is another one of those remarkable people whose intellectual curiosity knows no borders. He is also a bird watcher, and a darned good one. As I couldn't be in two places at the same time, I was desperate for help. When I invited him to run moth traps at other locales, he jumped at the opportunity.

Over the course of several years, Larry trapped peppered moths at three locations. The first was at Bays Mountain Park and Planetarium near Kingsport, Tennessee. They allowed him to use an observers' treehouse built in the middle of a wolf sanctuary. This made moth trapping all the more exciting for Larry, especially at night, when the wolves howled at the moon—or up at Larry's sleeping quarters. I went there only one time, to help Larry set up his traps. The next morning I drove to Pennsylvania as quickly as I could, to trap in the coal regions. No time to waste. And no wolves there.

Larry's second trapping location was in Weirton, West Virginia, about 30 miles west of Pittsburgh, Pennsylvania, once the center of the American steel industry. He was using mercury vapor lamp traps but catching very few moths. I was 300 miles to the east of Larry's site, running assembling traps—which put out no light and made no sound—on the culm banks surrounding abandoned coal mines. I was worried that I'd lose my generator to thieves if I left it unattended overnight. As I had a surplus of virgin females, I drove out to Weirton to provide Larry with an assembling trap. The next morning the place was dripping with male peppered moths. Larry couldn't believe it. Neither could I. Pheromones work! If we just could manufacture this species-specific pheromone, we could trap peppered moths wherever and whenever they were on the wing. But for us, at that time, we had to use real female peppered moths to send out their perfume.

The third and most important place Larry ran moth traps was back on the E. S. George Reserve in Michigan. I hadn't been there since 1995, when the frequency of melanics was 20%, based on a small capture of 35 moths. I was very curious to learn what had happened in six years. The latest reports from England at that time were that melanism had fallen to 5% at Caldy Common. Mention of this fact prompts me to include more sad news.

Somewhere in my files I have a short, handwritten aerogramme from Cyril, sent in 1998. It said, "Féo died today. We are devastated." Truly he was. As were all of us who knew Lady Clarke. I suspected Cyril could not carry on for long without her. His loyal assistants rallied to keep him, and his activities, going for as long as humanly possible. But it was too tall an order for them, or for anyone. There are a lot of stories about his decline into senescence; ultimately, he was put into a nursing home. I went to visit him there. He was very excited to see me. He told me he was expecting, that very day, a visitor from America. Gee, I wondered, who might that be? Then I realized *I* was that visitor. Yet through it all, there were moments of clarity. Lady Clarke had covered for him for the past several years, but now he was flying solo.

I met with his neighbor, a fellow professor who knew very little about moths, but he agreed to serve as the nominal supervisor of Cyril's highly capable assistants, in order to keep Cyril's moth trap going. They'd be paid from Cyril's Nuffield grant, but the university's administrators wouldn't allow the work of mere technicians to go unsupervised. So, officially we had a University of Liverpool professor from another department assume that role. And, for expert guidance, Laurence Cook at the University of Manchester agreed to look in on the operation. All of Cyril's technician's from my time at Caldy Common had retired, so the moth work fell to their replacement, Sally Thompson. She understood the operation inside and out, and she knew peppered moths as well as anyone. We all had complete confidence in Sally's scoring. She kept the operation going for two more years after Cyril died in 2000. When Larry Wiseman went back to the George Reserve in 2001, melanism at Caldy Common was down to 5%.

We piled into Larry's truck, along with several moth traps and gear, and headed to the George Reserve. This time we knew enough to arrive there as early as we could in May, so we wouldn't miss the first brood of the season—a lesson I'd learned the hard way. The plan was that I'd stay there for a night or two to help Larry get established, then he'd put me on a Greyhound bus heading back east, so I could run traps in Schuylkill County, Pennsylvania, as close to T. R. Manley's location as I could get.

Once on the George Reserve, we put traps near where Larry would be housed in the Camburn Lab. As it was well before dark, I suggested we take a walk around the reserve to get a feel for the place. After it got dark, we got lost. We found our way back by catching hold of the chain-link fence that encloses the entire reserve. It's roughly rectangular, so whichever way we chose to go, we'd eventually return to our starting point. But it's only *roughly* rectangular. There are a lot of ins and outs, and some of these go in or out for great distances. The reserve is a huge place, one I had never fully explored until that night—when I was showing Larry around. Some sections of the chain-link fence go over water. That includes small lakes, large ponds, and swamps. And some sections of the fence encroach on the neighboring backyards of local residents, who turned on floodlights when they heard us moving through the heavy brush outside their homes. That's when Larry grew nervous. Michigan rural folks might have guns, and maybe they don't like strangers crawling across their backyards, even if they are on the other side of a fence. Well, ultimately we made it back to Camburn Lab. We turned on the moth traps and had a beer.

Right away we had a pretty good catch. But no melanics. I took Larry to other places on the reserve to run traps. Outside my old garage residence, I instructed Larry to follow the road in order to reach the north gatehouse. "Are you sure?" he asked.

From our work, coupled with atmospheric records of relevant pollutants provided by governmental agencies in Virginia, Pennsylvania, and Michigan, we published a paper in 2002, titled "Recent history of melanism in American peppered moths," again in the *Journal of Heredity*. This time we didn't even ask for the cover, but the color photo of American peppered moths we submitted with our manuscript was chosen by the editors.

Larry was initially hesitant about putting his name on a paper with me, concerned that so many of my previous coauthors soon died. I assured him that publishing with me wasn't life threatening: Cyril Clarke lived to age 93, and Denis was a heavy smoker long before we met. I was not sure about Derek Whiteley's cause of death, but in any case, he wasn't a coauthor. Besides, Larry had already published a technical comment with me in *Science* in 1980, and he was still alive more than 20 years later. With that reassurance, he agreed to sign on. The important data were his, so I needed his consent, dead or alive.

What our paper confirmed is that the trends for the decline in melanism in the United States broadly correspond geographically to the nationwide trends in the United Kingdom. Lower numbers of melanic peppered moths in these locales are generally correlated with declines in atmospheric SO_2. Of particular interest is the continued close parallel between Caldy Common (near Liverpool, England) and the George Reserve (near Detroit, Michigan), both with levels that had dropped to around 5% by 2001. This time the frequency of melanic peppered moths at the George Reserve was estimated using a robust sample size of $N = 283$, thanks to Larry Wiseman. There is no question about the parallel evolution of melanic peppered moths, in two directions, on separate continents. It is a fact clearly supported by hard data.

Epilogue

Industrial melanism in peppered moths, because it is such a big story—the shining paradigm of observable evolution by natural selection—has long been a prime target of anti-evolutionists, fundamentalist creationists, intelligent-design creationists, and journalists eager for controversy. I don't want this book, intended to salute my colleagues, to end on a sour note. I've engaged in those battles elsewhere. Yet the war is hardly over (see a review of this issue by Laurence Cook and John Turner, published in 2020).

I prefer to ignore people with axes to grind or hidden agendas. I don't want to widen their audience by including them in discussions, as if they have legitimate criticisms to make. They don't. They behave more like unethical politicians than scientists. They are determined to win and deliberately distort the evidence supporting the science they seek to undermine. Headlines like "PEPPERED MOTHS DON'T REST ON TREE TRUNKS" are flat out wrong. Substantial data document that peppered moths commonly rest on the trunks of trees. Michael Majerus spent years climbing trees to find that out. Of the 135 peppered moths he saw and reported, 48 (35%) were on the trunks, 70 (52%) on branches, and 17 (13%) on twigs. So why would anyone make the false claim that peppered moths don't rest on tree trunks—or fail to retract such untruths in the light of hard evidence? Why cry "fraud," as if the whole story of industrial melanism is built on a house of cards and should be thrown out? This is the underhanded work of intelligent-design creationists who engage in intentional deception, a topic I discussed in a book review invited by *Skeptic* magazine in 2004.

Others have impugned the scientific integrity of Bernard Kettlewell, suggesting that he cooked his data to fit his arguments. There is no evidence supporting such serious charges, and I have dealt with that in a book review invited by *Science* in 2002.

The sad fact is that the steady drumbeat of unfounded criticism targeting industrial melanism has taken its toll. Some years ago, one of my undergraduate research students was interviewing for admission to graduate school at a prestigious university and had a conversation with a highly energetic geneticist, whose name I'll withhold because he's now dead and can't defend himself. He asked my student, "So, what's Bruce Grant up to these days?"

"He's working on peppered moths," said my student.

"Oh, that story is so flawed," said the famous professor.

When my student reported this back to me, I was flabbergasted. "Really? In what way is it flawed? Did he say? Did you ask him?"

My student's experience was not unique. It's just an example. Too many people have been swayed by disinformation, and it has come at a sizeable cost to science education.

I am not suggesting that this research on peppered moths, or any other work, should be protected from criticism. It is very much part of our job as scientists to examine—scrutinize—every detail of every claim, not only to ensure that conclusions are supported by evidence, but also to question *how* that evidence was obtained. Moreover, we must be prepared to amend our conclusions in light of new evidence, as these assessments are always conditional and subject to change. This is not a weakness of science—it is its strength.

We don't settle issues through endless debate or, worse, by deliberately misrepresenting evidence. In science, the correctness of any work is determined by its repeatability. And that's exactly what happened in addressing recognized deficiencies that were legitimately leveled at the bird predation studies of Kettlewell and his successors.

By far the largest of these experiments were those conducted by Michael Majerus from 2001 through 2006. Unfortunately, Majerus died before he could formally publish that research. He did, however, present it at a meeting in Uppsala, Sweden, and posted the contents online. Not long after his death, the librarian at the University of Cambridge removed Majerus's posting, to conform with some institutional or library policy that made no sense to me when I enquired about it. Several of my colleagues and I thought this work should be available for all to see, so we decided to publish it in *Biology Letters* in 2012, with the title "Selective predation on the peppered moth: the last experiment of Michael Majerus." Since he was deceased, Majerus could not be listed among the coauthors (Cook, Grant, Saccheri, and Mallet), so we put his name up front, in the title. There is no mistaking whose work this was. It took four of us to pull all of it together, reanalyze the data, make graphs, get background information not expressly included in the Uppsala talk, and write it up. The full details are available in the article itself. Here is the abstract of that paper:

> Colour variation in the peppered moth *Biston betularia* was long accepted to be under strong natural selection. Melanics were believed to be fitter than pale morphs because of lower predation at daytime resting sites on dark, sooty bark. Melanics became common during the industrial revolution, but since 1970 there has been a rapid reversal, assumed to have been caused by predators selecting against melanics resting on today's less sooty bark. Recently, these classical explanations of melanism were attacked, and there has been general scepticism about birds as selective agents. Experiments and observations were accordingly carried out by Michael Majerus to address perceived weaknesses of earlier work. Unfortunately, he did not live to publish the results, which are analysed and presented here by the authors. Majerus released 4864 moths in his six-year experiment, the largest ever attempted for any similar study. There was strong differential bird predation against melanic peppered moths. Daily selection against melanics ($s \simeq 0.1$) was sufficient in magnitude and direction to explain the recent rapid decline of melanism in post-industrial Britain. These data provide the most direct evidence yet to implicate camouflage and bird predation as the overriding explanation for the rise and fall of melanism in moths.

Not long after its publication, the editor of *Biology Letters*, Charlotte Wray, sent us a congratulatory note that included the following: "I am delighted to inform you that your paper 'Selective bird predation on the peppered moth: the last experiment of Michael Majerus' has recently been included in the *Biology Letters* most downloaded and cited articles from 2012. Your paper was one of our top ten download articles in 2012 **AND** [emphasis in original] a top cited article."

Just suppose, for the sake of argument, that (1) Majerus hadn't done those experiments, (2) we decided to ignore all of Kettlewell's work on the subject of predation, and (3) we were unaware of all the other work on predation done by Kettlewell's successors and contemporaries. What would be left?

We would still have the largest continuous record of allele frequency changes ever recorded in populations at multiple locations by numerous people. No force known to science can account for the velocity and trajectory of these changes, other than natural selection. So even if bird predation were not the proximal agent for that selection, obviously something is driving this evolution.

One must clearly distinguish between the *evidence* for natural selection and the *mechanism* of that selection. In an analogous illustration, a dead body with a knife in its back is evidence that a murder has been committed. A failure to establish beyond reasonable doubt the guilt of a leading suspect does not mean that the murder did not occur. Yet critics of the predation experiments consistently conflate problems in those experiments with evidence for natural selection. The evidence for natural selection does not depend on establishing bird predation as the proximal agent. It is based on the observed changes in allele frequencies in peppered moth populations—and these data are indisputable.

Now that we have firmly established that the evidence for natural selection comes from population data, we may return to the question of the proximal selection mechanism. What is it? Bird predation! Majerus directly observed and identified nine different bird species that consumed peppered moth phenotypes selectively. The weight of evidence supporting bird predation as the primary agent driving the evolution of melanism in *Biston betularia* is now substantial. It has not been contradicted, either by experiment or observation, after a half century of intensive reexamination.

Our discovery of the parallel evolution of melanism in both directions—its rise and subsequent fall—in American peppered moths included no direct assessment of predation. What we did show is that changes in melanic frequencies occurred in concert with directional changes in atmospheric concentrations of suspended particulates and SO_2. We picked these particular pollutants because they have been those reported by British workers as the ones most strongly correlated geographically with observed changes in the percentages of melanic moths in various regions of Britain. Investigators there have shown that reflectance from the surface of tree bark is strongly negatively correlated with atmospheric levels of suspended particles, and that lichens are vulnerable to high levels of atmospheric SO_2.

While a correlation alone does not establish a causal relationship, common correlations suggest a common cause.

| Bibliography

Asami T and Grant BS (1995) Melanism has not evolved in Japanese *Biston betularia* (Geometridae). Journal of the Lepidopterists' Society 49: 88–91.

Berry RJ (1990) Industrial melanism and peppered moths (*Biston betularia* L.). Biological Journal of the Linnean Society 39: 301–322.

Bishop JA (1972) An experimental study of the cline of industrial melanism in *Biston betularia* (L.) (Lepidoptera) between urban Liverpool and rural North Wales. Journal of Animal Ecology 41:209–243.

Bishop JA, Cook LM, and Muggleton J (1978) The response of two species of moths to industrialization in northwest England I: polymorphisms for melanism. Philosophical Transactions of the Royal Society of London B, 281: 491–515.

Boardman M, Askew RR, and Cook LM (1974) Experiments on resting-site selection by nocturnal moths. Journal of Zoology 172: 343–355.

Bourguet D (1999) The evolution of dominance. Heredity 83: 1–4.

Brakefield PM and Liebert TG (1990) The reliability of estimates of migration in the peppered moth *Biston betularia* and some implications for selection-migration models. Biological Journal of the Linnean Society 39: 335–341.

Brakefield PM and Liebert TG (2000) Evolutionary dynamics of declining melanism in the peppered moth in the Netherlands. Proceedings of the Royal Society B 267: 1953–1957.

Clarke CA, Clarke FMM, and Grant B (1993) *Biston betularia* (Geometridae), the peppered moth, in Wirral, England: an experiment in assembling. Journal of the Lepidopterists' Society 47: 17–21.

Clarke CA, Clarke FMM, and Owen DF (1991) Natural selection and the scarlet tiger moth, *Panaxia dominula*: inconsistencies in the scoring of the heterozygote, f. *medionigra*. Proceedings of the Royal Society B 244: 203–205.

Clarke CA, Grant B, Clarke FMM, and Asami T (1994) A long term assessment of *Biston betularia* (L.) in one UK locality (Caldy Common near West Kirby, Wirral) 1959–1993, and glimpses elsewhere. Linnean 10: 18–26.

Clarke CA, Mani GS, and Wynne G (1985) Evolution in reverse: clean air and the peppered moth. Biological Journal of the Linnean Society 26: 189–199.

Clarke CA and Sheppard PA (1966) A local survey of the distribution of the industrial melanic forms of the moth *Biston betularia* and estimates of the selective values of these forms in an industrial environment. Proceedings of the Royal Society B 165: 424–439.

Cook LM (2000) Changing views on melanic moths. Biological Journal of the Linnean Society 69: 431–441.

Cook LM (2003) The rise and fall of the *carbonaria* form of the peppered moth. Quarterly Review of Biology 78: 399–417.

Cook LM, Dennis RHL, and Mani GS (1999) Melanic morph frequency in the peppered moth in the Manchester area. Proceedings of the Royal Society B 266: 293–297.

Cook LM and Grant BS (2000) Frequency of *insularia* during the decline in melanics in the peppered moth *Biston betularia* in Britain. Heredity 85: 580–585.

Cook LM, Grant BS, Saccheri IJ, and Mallet J (2012) Selective bird predation on the peppered moth: the last experiment of Michael Majerus. Biology Letters, https://doi.org/10.1098/rsbl.2011.1136/.

Cook LM, Mani GS, and Varley ME (1986) Postindustrial melanism in the peppered moth. Science 231: 611–613.

Cook LM, Sutton SL, and Crawford TJ (2005) Melanic moth frequencies in Yorkshire, an old English industrial hotspot. Journal of Heredity 96: 522–528.

Cook LM and Turner JRG (2008) Decline of melanism in two British moths: spatial, temporal and interspecific variation. Heredity 101: 483–489.

Cook LM and Turner JRG (2020) Fifty per cent and all that: what Haldane actually said. Biological Journal of the Linnean Society 129: 765–771.

Creed ER, Lees DR, and Bulmer MG (1980) Pre-adult viability differences of melanic *Biston betularia* (L.) (Lepidoptera). Biological Journal of the Linnean Society 13: 251–262.

Creed ER, Lees DR, and Duckett JG (1973) Biological method of estimating smoke and sulphur dioxide pollution. Nature 244: 278–280.

Cuthill IC, Partridge JC, Bennett ATD, Church SC, Hart NS, and Hunt S (2000) Ultraviolet vision in birds. Advances in the Study of Behaviour 29: 159–214.

Douwes P, Mikkola K, Petersen B, and Vestergren A (1976) Melanism in *Biston betularius* from north-west Europe (Lepidoptera: Geometridae). Entomologica Scandinavia 7: 261–266.

Eacock A, Rowland HM, van't Hof AE, Yung CJ, Edmonds N, and Saccheri IJ (2019) Adaptive color change and background choice in peppered moth caterpillars is mediated by extraocular photoreception. Communications Biology 2: 286.

Edleston RS (1864) 88: *Amphydasis* [=*Amphidasis*] *betularia*. Entomologist 2: 150.

Endler JA (1986) Natural Selection in the Wild. Princeton, NJ: Princeton University Press.

Ford EB (1975) Ecological Genetics, 4th ed. London: Chapman & Hall.

Grant BS (1999) Fine tuning the peppered moth paradigm. Evolution 53: 980–984.

Grant BS (2002) Sour grapes of wrath. Science 297: 940–941.

Grant BS (2004) Allelic melanism in American and British peppered moths. Journal of Heredity 95: 97–102.

Grant BS (2004) Intentional deception. Skeptic 11: 84–86.

Grant BS (2012) Industrial melanism. eLS: Citable Reviews in the Life Sciences, doi:10.1002/9780470015902.a0001788.pub3.

Grant BS and Clarke CA (1999) An examination of intraseasonal variation in the incidence of melanism in peppered moths, *Biston betularia* (Geometridae). Journal of the Lepidopterists' Society 53: 99–103.

Grant BS, Cook AD, Owen DF, and Clarke CA (1998) Geographic and temporal variation in the incidence of melanism in peppered moth populations in America and Britain. Journal of Heredity 89: 465–471.

Grant BS and Howlett RJ (1988) Background selection by the peppered moth (*Biston betularia* Linn.): individual differences. Biological Journal of the Linnean Society 33: 217–232.

Grant BS, Owen DF, and Clarke CA (1995) Decline of melanic moths. Nature 373: 565.

Grant BS, Owen DF, and Clarke CA (1996) Parallel rise and fall of melanic peppered moths in America and Britain. Journal of Heredity 87: 351–357.

Grant BS and Wiseman LL (2002) Recent history of melanism in American peppered moths. Journal of Heredity 93: 86–90.

Haldane JBS (1924) A mathematical theory of natural and artificial selection. Transactions of the Cambridge Philosophical Society 23: 19–41.

Harrison JWH (1927) The induction of melanism in the Lepidoptera and its evolutionary significance. Nature 119: 127–129.

Hooper J (2002) Of Moths and Men: The Untold Story of Science and the Peppered Moth. New York: W. W. Norton.

Howlett RJ and Majerus MEN (1987) The understanding of industrial melanism in the peppered moth (*Biston betularia*) (Lepidoptera). Biological Journal of the Linnean Society 30: 31–34.

Kettlewell HBD (1955) Selection experiments on industrial melanism in the Lepidoptera. Heredity 9: 323–342.

Kettlewell HBD (1956) Further selection experiments on industrial melanism in the Lepidoptera. Heredity 10: 287–301.

Kettlewell [H]B[D] (1973) The Evolution of Melanism: The Study of Recurring Necessity. Oxford: Clarendon Press.

Lees DR (1981) Industrial melanism: genetic adaptation of animals to air pollution. In: Bishop JA and Cook LM (eds.) Genetic Consequences of Man-Made Change, pp. 129–176. New York: Academic Press.

Lees DR and Creed ER (1977) The genetics of the *insularia* forms of the peppered moth, *Biston betularia*. Heredity 39: 67–73.

Leverton R (2001) Enjoying Moths. London: Academic Press.

Liebert TG and Brakefield PM (1987) Behavioural studies on the peppered moth *Biston betularia* and a discussion of the role of pollution and lichens in industrial melanism. Biological Journal of the Linnean Society 31: 129–150.

Majerus MEN (1998) Melanism: Evolution in Action. Oxford: Oxford University Press.

Majerus MEN (2008) Non-morph specific predation on peppered moths (*Biston betularia*) by bats. Ecological Entomology 33: 679–683.

Majerus MEN (2009) Industrial melanism in the peppered moth, *Biston betularia*: an excellent teaching example of Darwinian evolution in action. Evolution: Education and Outreach 2: 63–74, https://doi.org/10.1007/s12052–008–0107-y/.

Majerus MEN, Brunton CFA, and Stalker J (2000) A bird's eye view of the peppered moth. Journal of Evolutionary Biology 13: 155–159.

Mallet J (2004) The peppered moth: a black and white story after all. Genetics Society News 50: 34–38.

Mani GS (1982) A theoretical analysis of the morph frequency variation in the peppered moth over England and Wales. Biological Journal of the Linnean Society 17: 259–267.

Manley TR (1988) Temporal trends in frequencies of melanic morphs in cryptic moths of rural Pennsylvania. Journal of the Lepidopterists' Society 42: 213–217.

Marrs RH, Hicks MJ, and Fuller RM (1986) Losses of lowland heath through succession at four sites in Breckland, East Anglia, England. Biological Conservation 36: 19–38.

Mettler LE, Gregg TG, and Schaffer HE (1988) Population Genetics and Evolution, 2nd ed. Englewood Cliffs, NJ: Prentice-Hall.

Mikkola K (1984) Dominance relations among the melanic forms of *Biston betularius* and *Odontopera bidentata* (Lepidoptera: Geometridae) Heredity 52: 9–16.

Mikkola K (1984) On selective forces acting in the industrial melanism of *Biston* and *Oligia* moths (Lepidoptera: Geometridae and Noctuidae). Biological Journal of the Linnean Society 21: 409–421.

Mikkola K and Rantala MJ (2010) Immune defence, a possible nonvisual selective factor behind the industrial melanism of moths (Lepidoptera). Biological Journal of the Linnean Society 99: 831–838.

Noor MAF, Parnell RS, and Grant BS (2008) A reversible color polyphenism in American peppered moth (*Biston betularia cognataria*) caterpillars. PLoS ONE 3(9): e3142, doi: 10.1371/journal.pone3.0003142.

Owen DF (1961) Industrial melanism in North American moths. American Naturalist 95: 227–233.

Owen DF (1962) The evolution of melanism in six species of North American geometrid moths. Annals of the Entomological Society of America 55: 695–703.

Owen DF (1962) Parallel evolution in European and North American populations of a geometrid moth. Nature 195: 830–831.

Owen DF (1997) Natural selection and evolution in moths: homage to J. W. Tutt. Oikos 78: 177–181.

Owen DF and Clarke CA (1993) The *medionigra* polymorphism, *Panaxia dominula* (Lepidoptera: Artiidae): a critical re-assessment. Oikos 67:393–402.

Rindge FH (1975) A revision of the New World Bistonini (Lepidoptera: Geometridae). Bulletin of the American Museum of Natural History 156: 69–155.

Saccheri IJ, Rousett F, Watts PC, Brakefield PM, and Cook LM (2008) Selection and gene flow on a diminishing cline of melanic peppered moths. Proceedings of the National Academy of Sciences of the USA 105: 16212–16217.

Sargent TD (1966) Background selections of geometrid and noctuid moths. Science 154: 1674–1675.

Sargent TD (1968) Cryptic moths: effects on background selections of painting the circumocular scales. Science 159: 100–101.

Sargent TD (1969) Background selections of the pale and melanic forms of the cryptic moth *Phigalia titea* (Cramer). Nature 222: 585–586.

Sargent TD, Millar CD, and Lambert DM (1998) The "classical" explanation of industrial melanism. Evolutionary Biology 30: 299–322.

Scudder GGE (1972) Industrial melanism: a possibility in British Columbia. Journal of the Entomological Society of British Columbia 69: 46–48.

Seaward MRD (1993) Lichens and sulphur dioxide air pollution: field studies. Environmental Reviews 1: 73–91.

Skelhorn J, Rowland HM, Speed MP, and Ruxton GD (2010) Masquerade: camouflage without crypsis. Science 327: 51.

Steward RC (1977) Industrial and non-industrial melanism in the peppered moth, *Biston betularia* (L.). Ecological Entomology 2: 231–243.

Tutt JW (1896) British Moths. London: Routledge.

van't Hof AE, Campagne P, Rigden DJ, Yung CJ, Lingley J, Quail MA, Hall N, Darby AC, and Saccheri IJ (2016) The industrial melanism mutation in British peppered moths is a transposable element. Nature 534: 102–105, https://doi.org/10.1038/nature17951/.

van't Hof AE, Edmonds N, Dalikova M, Marec F, and Saccheri IJ (2011) Industrial melanism in British peppered moths has singular and recent mutational origin. Science 332: 958–960.

West DA (1977) Melanism in *Biston* (Lepidoptera: Geometridae) in the rural central Appalachians. Heredity 39: 75–81.

Yoon CK (1996) Parallel plots in classic of evolution. New York Times, Science Times, Nov. 12, pp. C1 and C7, https://www.nytimes.com/1996/11/12/science/parallel-plots-in-classic-of-evolution.html.

Young M and Musgrave I (2005) Moonshine: why the peppered moth remains an icon of evolution. Skeptical Inquirer 29: 23–28.

Index